PRELIMINARY CONCEPTS

Volume 10: Analytic Properties of Trigonometric Functions

All variables denote real numbers unless otherwise noted.

Symbols

α°	α degrees
α^{R}	α radians
sin α, csc α	$\sin\alpha = \frac{y}{r}$ $\quad \csc\alpha = \frac{r}{y} \quad (y \neq 0)$
cos α, sec α	$\cos\alpha = \frac{x}{r}$ $\quad \sec\alpha = \frac{r}{x} \quad (x \neq 0)$
tan α, cot α	$\tan\alpha = \frac{y}{x} \quad (x \neq 0)$ $\quad \cot\alpha = \frac{x}{y} \quad (y \neq 0)$
A	{all angles}
$\sin^2\alpha, \cos^2\alpha$, **etc.**	$(\sin\alpha)^2, (\cos\alpha)^2$, etc.
α'	α minutes
$\tilde{\alpha}$	Reference angle for angle α in standard position

$r = \sqrt{x^2 + y^2}$

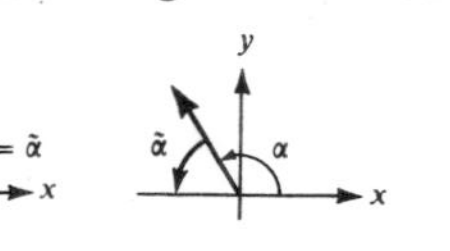

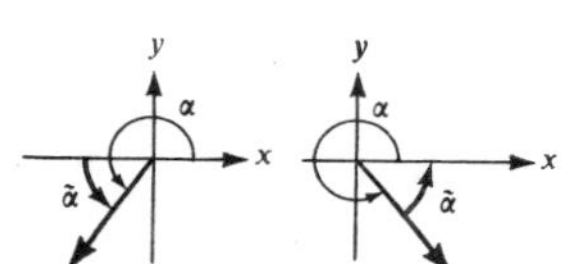

A,B,C	The angles of a triangle
a,b,c	The lengths of sides of a triangle opposite angles *A*, *B*, and *C*, respectively

m(α)	Measure of angle α
m°(α)	Degree measure of α
m^R(α)	Radian measure of α

Properties

Conversion formulas for angle measure:

$$\alpha^{\circ} = \frac{180}{\pi}\alpha^{R}; \quad \alpha^{R} = \frac{\pi}{180}\alpha^{\circ}$$

Coterminal angles:

$$\alpha^{\circ} \text{ and } \alpha^{\circ} + k \cdot 360^{\circ} \quad \text{or} \quad \alpha^{R} \text{ and } \alpha^{R} + k \cdot 2\pi^{R}, \quad k \in J$$

are measures of coterminal angles.

Domains and ranges of the trigonometric functions:

Function	*Domain*	*Range*
sine *cosine*	$\{\alpha \mid \alpha \in A\}$	$-1 \le \sin\alpha \le 1$ $-1 \le \cos\alpha \le 1$
tangent *secant*	$\{\alpha \mid \alpha \in A, \alpha \neq 90^{\circ} + k \cdot 180^{\circ}, k \in J\}$	$-\infty < \tan\alpha < \infty$ $\sec\alpha \le -1$ or $\sec\alpha \ge 1$
cotangent *cosecant*	$\{\alpha \mid \alpha \in A, \alpha \neq k \cdot 180^{\circ}, k \in J\}$	$-\infty < \cot\alpha < \infty$ $\csc\alpha \le -1$ or $\csc\alpha \ge 1$

Properties *(continued)*

Signs for the trigonometric functions:

Quadrant	I $x,y > 0$	II $x < 0, y > 0$	III $x,y < 0$	IV $x > 0, y < 0$
$\sin\alpha$ *or* $\csc\alpha$	+	+	−	−
$\cos\alpha$ *or* $\sec\alpha$	+	−	−	+
$\tan\alpha$ *or* $\cot\alpha$	+	−	+	−

Function values for special angles:

α°	α^R	$\sin\alpha$	$\cos\alpha$	$\tan\alpha$	$\csc\alpha$	$\sec\alpha$	$\cot\alpha$
0°	0^R	0	1	0	*Undefined*	1	*Undefined*
30°	$\frac{\pi^R}{6}$	$\frac{1}{2}$	$\frac{\sqrt{3}}{2}$	$\frac{1}{\sqrt{3}}$	2	$\frac{2}{\sqrt{3}}$	$\sqrt{3}$
45°	$\frac{\pi^R}{4}$	$\frac{1}{\sqrt{2}}$	$\frac{1}{\sqrt{2}}$	1	$\sqrt{2}$	$\sqrt{2}$	1
60°	$\frac{\pi^R}{3}$	$\frac{\sqrt{3}}{2}$	$\frac{1}{2}$	$\sqrt{3}$	$\frac{2}{\sqrt{3}}$	2	$\frac{1}{\sqrt{3}}$
90°	$\frac{\pi^R}{2}$	1	0	*Undefined*	1	*Undefined*	0
180°	π^R	0	−1	0	*Undefined*	−1	*Undefined*
270°	$\frac{3\pi^R}{2}$	−1	0	*Undefined*	−1	*Undefined*	0

Some useful approximations:

$$\sqrt{2} \approx 1.414, \quad \frac{1}{\sqrt{2}} \approx 0.707, \quad \sqrt{3} \approx 1.732$$

$$\frac{\sqrt{3}}{2} \approx 0.866, \quad \frac{1}{\sqrt{3}} \approx 0.577, \quad \frac{2}{\sqrt{3}} \approx 1.155$$

Formulas for using the reference angle $\tilde{\alpha}$:

When $\text{trig}\,\alpha \geq 0$, $\text{trig}\,\alpha = \text{trig}\,\tilde{\alpha}$

When $\text{trig}\,\alpha < 0$, $\text{trig}\,\alpha = -\text{trig}\,\tilde{\alpha}$

Trigonometric ratios in a right triangle:

$$\sin\alpha = \frac{\text{length of side opposite } \alpha}{\text{length of hypotenuse}}$$

$$\cos\alpha = \frac{\text{length of side adjacent to } \alpha}{\text{length of hypotenuse}}$$

$$\tan\alpha = \frac{\text{length of side opposite } \alpha}{\text{length of side adjacent to } \alpha}$$

$$\csc\alpha = \frac{\text{length of hypotenuse}}{\text{length of side opposite } \alpha}$$

$$\sec\alpha = \frac{\text{length of hypotenuse}}{\text{length of side adjacent to } \alpha}$$

$$\cot\alpha = \frac{\text{length of side adjacent to } \alpha}{\text{length of side opposite } \alpha}$$

Properties (*continued*)

Solution of oblique triangles with sides of length a, b, *and* c *opposite angles* α, β, *and* γ *respectively*:

Given conditions	Property to be used
1. *Two angles and the length of one side*	*Law of sines:* $\frac{\sin\alpha}{a} = \frac{\sin\beta}{b} = \frac{\sin\gamma}{c}$ *or* $\frac{a}{\sin\alpha} = \frac{b}{\sin\beta} = \frac{c}{\sin\gamma}$
2. *Lengths of two sides and an angle opposite one of the sides; possibilities:* **a.** *No triangle* **b.** *One right triangle* **c.** *One oblique triangle* **d.** *Two oblique triangles (ambiguous case)*	
3. *Lengths of two sides and the included angle*	*Law of cosines:* $c^2 = a^2 + b^2 - 2ab\cos\gamma$, $b^2 = a^2 + c^2 - 2ac\cos\beta$, $a^2 = b^2 + c^2 - 2bc\cos\alpha$
4. *Lengths of three sides*	

Circular motion:

Length of circular arc:

$$s = r \cdot m^R(\alpha)$$

Linear velocity:

$$v = \frac{r \cdot m^R(\alpha)}{t}$$

$$v = r\omega \quad (\omega \text{ in radians per unit time})$$

ANALYTIC PROPERTIES OF TRIGONOMETRIC FUNCTIONS

BERNARD FELDMAN

Los Angeles Pierce College

Consulting Editor

Irving Drooyan

Los Angeles Pierce College

Wadsworth Publishing Company, In
Belmont, California

Designers: Wadsworth Design Staff

Mathematics Editor: Don Dellen

Production Editor: Dorothy Graham

Copy Editor: Alice Goehring

Technical Illustrator: Carleton Brown

ISBN 0-534-00357-5
L. C. Cat. Card No. 74-78036
Printed in the United States of America

1 2 3 4 5 6 7 8 9 10—78 77 76 75 74

PREFACE

About the series

This is the tenth volume in the Wadsworth Precalculus Mathematics Series for college students. The titles of the twelve volumes in the series are listed on the back cover. Selected volumes can be used (1) as the basic textbook for any standard precalculus course, (2) for independent study as a supplement to another text used in the course, or (3) for a self-study review of material previously studied by a student.

The modular organization and following features of this series make it particularly useful for independent work or to accommodate individual differences in formal courses.

1. Beginning with the second volume, each volume includes lists of all the mathematical symbols and properties covered in previous volumes that are necessary prerequisites for that particular volume.
2. The textual material in each section is presented in a concise style to highlight the basic mathematical ideas and instructional objectives of the section.
3. Each group of exercises covering a particular topic is preceded by detailed examples of a similar type.
4. A detailed Solution Key for all exercises is included in volumes 1 to 8. Solution Keys for the odd-numbered exercises are included in volumes 9 to 12.

In formal lecture presentations, it is anticipated that instructors will indicate how the material in a particular section relates to the preceding and subsequent topics in the volume, and that they will introduce mathematical structure, from elementary heuristic motivations to an axiomatic approach, commensurate with the objectives of the course. Instructors may wish to make a distinction between axioms and theorems, which we have listed together as properties. Also, they may wish to prove some of the theorems. Because of the detailed examples and Solution Key contained in each volume, lectures for each section can be brief. Hence, if facilities are available, these lectures are readily adaptable to short videotape presentations.

The list of volumes in this series presented on the back cover shows the number of sections and the number of review exercise sets (unit reviews) in each volume. Since each section and the review exercise set at the end of each unit have been designed as an assignment for one class period, this information can be helpful in selecting the number of volumes that can be completed for courses of different lengths.

About this volume

This volume develops the notion of circular functions which pair real numbers with real numbers and is concerned with analytic aspects of both trigonometric and circular functions that are useful in more advanced courses. Introductory topics of complex numbers, which are prerequisite for the material in Unit 3, are included in the Appendix for students who lack this background.

The instructional objectives of each unit and the Appendix can be readily observed from three sources: the table of contents; the symbols and properties listed on the back endpapers, which summarize the topics introduced in this volume; and the review exercises at the end of each unit and the Appendix.

Volumes 9 and 10 together constitute a complete course in trigonometry. The front endpapers of this volume contain a list of the preliminary concepts of trigonometry from Volume 9.

Tables of squares and square roots, values of trigonometric functions, and values of circular functions that are needed in this volume are included.

Solutions are given only to odd-numbered exercises in each unit in order to keep this volume to a convenient size. However, the solutions to the even-numbered exercises are included in the Supplement to the series which is available to instructors.

The following preface to students includes our suggestions regarding effective use of this text.

BERNARD FELDMAN

TO THE STUDENT

The organization of the material in this volume probably differs from that of other mathematics textbooks you may have used. Therefore, we shall make some suggestions as to how to use this volume, whether you are using it as the basic textbook for your course, for independent study as a supplement to another text used in your class, or for review of material that you have studied previously.

To begin, acquaint yourself with the overall organization of this volume:

1. Mathematical symbols and properties that are prerequisites for the topics introduced in this volume are listed on the front endpaper. If many of these symbols and properties appear totally unfamiliar to you, you should probably review previous volumes in this series before beginning this one.
2. A Solution Key containing a complete solution for each odd-numbered exercise is included at the end of the volume.
3. Each section begins with a list of the definitions and notation and a list of the properties that are introduced and used in that section. The properties include both the assumptions that are made about numbers and operations and some logical consequences of these assumptions.
4. Each unit concludes with a set of review exercises which are representative of the exercises throughout the unit. This review may conveniently be used as a self-test for the unit or as a diagnostic test to determine if you need to study a particular unit.
5. A summary of the symbols and properties introduced in this volume is included on the back endpapers.

After you have become familiar with the organization of the material, we suggest that before starting a unit, you determine the instructional objectives you expect to have attained upon completing it. These objectives can be determined from the references noted in 3, 4, and 5 above.

When beginning each section, it is suggested that you give the introductory material a preliminary reading. It is not expected that the definitions, symbols, and properties will be very meaningful at this first reading. As you proceed to work the exercises, you will probably want to refer to this material frequently. After you complete the assignment, you should reread the introductory material, this time with greater understanding. The detailed examples in each set of exercises should be studied carefully before proceeding to the exercises that follow. The format presented in these examples should be closely followed in your work. As you work the exercises, you may wish to refer to the symbols and properties listed on the front endpaper (which we assume you have encountered in a previous course). However, no doubt you have forgotten some of these, and probably you will need to refer to this list frequently as you begin this volume and less frequently as you proceed through it.

The Solution Key can be used in various ways. You may want to check your answer to each odd-numbered exercise as it is completed, or perhaps you would rather check the answers after you have completed an entire exercise set. Remember that many exercises can be solved in different ways. In order to keep the Solution Key to a convenient size, alternative solutions are not shown. We anticipate that sometimes you will devise ways of solving problems that are simpler than ours. However, if you have difficulty with a particular exercise, you should refer to the solution given in the Solution Key to see one way in which the exercise may be solved. (Of course, it is best to do this only after you have made an honest attempt to work the problem yourself.) As previously suggested, upon completing a unit, you might want to evaluate your progress by considering the Unit Review as a self-test. If you do so, you should complete the entire review before checking your answers. The results (you may wish to give yourself a grade) should suggest whether you are ready to proceed to the next unit or whether you should review a particular section or sections before proceeding.

When you have completed this volume, you may want to detach the summary pages of symbols and properties on the back endpapers as a convenient reference chart for your further studies.

This volume has been prepared in such a way as to make *you* responsible for your success. Good luck!

BERNARD FELDMAN

CONTENTS

UNIT 1

PERIODIC PROPERTIES OF TRIGONOMETRIC FUNCTIONS

In addition to using trigonometric functions to solve triangles, their periodic properties enable us to use the equations which define such functions as mathematical models for a wide variety of periodic phenomena. In this unit we will study these properties with special emphasis on sketching the graphs of the functions.

1.1 CIRCULAR FUNCTIONS

Definitions and Notation

x The directed length of the arc intercepted on a unit circle by a central angle α; x is the measure of α in radians, $x = m^R(\alpha)$*

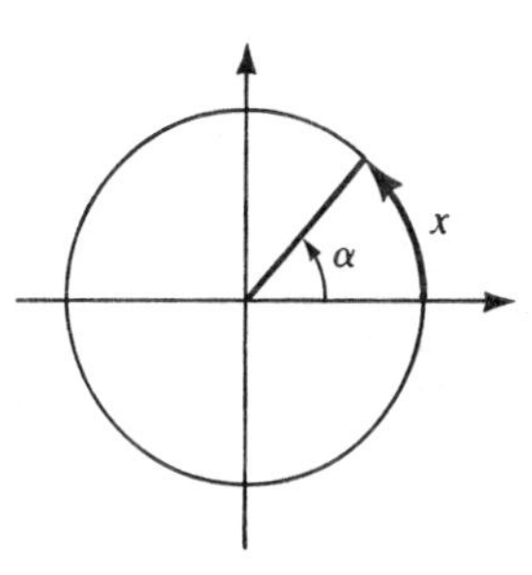

Examples

a.

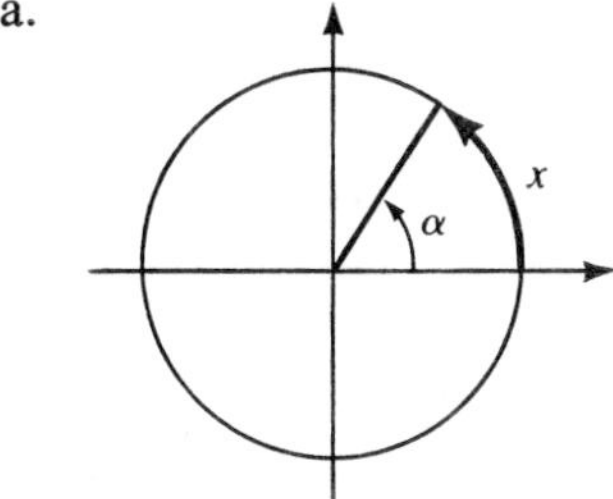

If $m^R(\alpha) = \frac{\pi}{3}$, then

$$x = \frac{\pi}{3} \approx \frac{3.14}{3} \approx 1.05$$

* Henceforth in this volume x and y will denote real numbers and α, β, γ, and θ will denote angles.

b. 

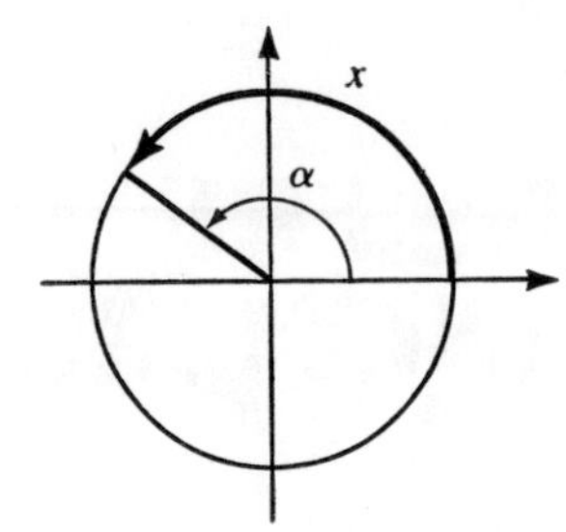

If $m^R(\alpha) = \frac{5\pi}{6}$, then

$$x = \frac{5\pi}{6} \approx \frac{5(3.14)}{6} \approx 2.62$$

sin x $\quad \sin x = \sin \alpha^R$

cos x $\quad \cos x = \cos \alpha^R$

tan x $\quad \tan x = \tan \alpha^R$

csc x $\quad \csc x = \csc \alpha^R$

sec x $\quad \sec x = \sec \alpha^R$

cot x $\quad \cot x = \cot \alpha^R$

Examples

a. $\cos \frac{\pi}{3} = \cos \frac{\pi^R}{3} = 0.5000$

b. $\tan \frac{5\pi}{6} = \tan \frac{5\pi^R}{6} = -\frac{1}{\sqrt{3}}$
≈ -0.5774

c. From Table III,
$\sin 0.25 = \sin 0.25^R \approx 0.2474$

d. From Table III,
$\sec 0.88 = \sec 0.88^R \approx 1.569$

Circular functions

Function	Domain
sine $= \{(x, \sin x)\}$ cosine $= \{(x, \cos x)\}$	R
tangent $= \{(x, \tan x)\}$ secant $= \{(x, \sec x)\}$	$x \neq \frac{\pi}{2} + k \cdot \pi \quad (k \in J)$
cotangent $= \{(x, \cot x)\}$ cosecant $= \{(x, \csc x)\}$	$x \neq k \cdot \pi \quad (k \in J)$

trig x $\quad \text{trig } x \in \{\sin x, \cos x, \tan x, \csc x, \sec x, \cot x\}$

Example

$$\text{trig} \frac{\pi}{6} \in \left\{\sin \frac{\pi}{6}, \cos \frac{\pi}{6}, \tan \frac{\pi}{6}, \csc \frac{\pi}{6}, \sec \frac{\pi}{6}, \cot \frac{\pi}{6}\right\}$$

trig α $\quad \text{trig } \alpha \in \{\sin \alpha, \cos \alpha, \tan \alpha, \csc \alpha, \sec \alpha, \cot \alpha\}$

Examples

a. $\text{trig}\,\frac{\pi^R}{6} \in \left\{\sin\frac{\pi^R}{6}, \cos\frac{\pi^R}{6}, \tan\frac{\pi^R}{6}, \csc\frac{\pi^R}{6}, \sec\frac{\pi^R}{6}, \cot\frac{\pi^R}{6}\right\}$

b. $\text{trig } 30° \in \{\sin 30°, \cos 30°, \tan 30°, \csc 30°, \sec 30°, \cot 30°\}$

Periodic function A function f for which there exists an $a > 0$ such that for all x in the domain, $f(x + a) = f(x), \quad a \in R$

Period of a function The smallest positive number a for a periodic function such that $f(x + a) = f(x), \quad a \in R$

Example

From a property of the trigonometric functions (see the Preliminary Concepts), we know that for every x, $\sin (x + 2\pi)^R = \sin x^R$, and it can be shown that 2π is the smallest number for which this is true. Thus the sine function is periodic with period 2π.

Reference arc The arc on a unit circle between the terminal point of an arc with length x and the horizontal axis

$\tilde{x}$ Length of the reference arc, $0 < \tilde{x} < \frac{\pi}{2}$

Examples

a.

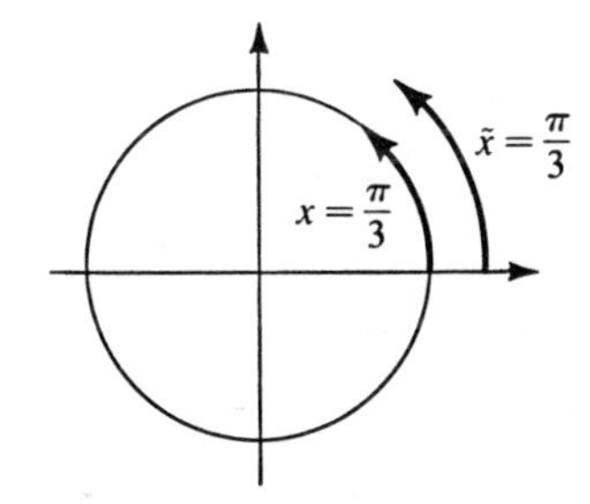

b.

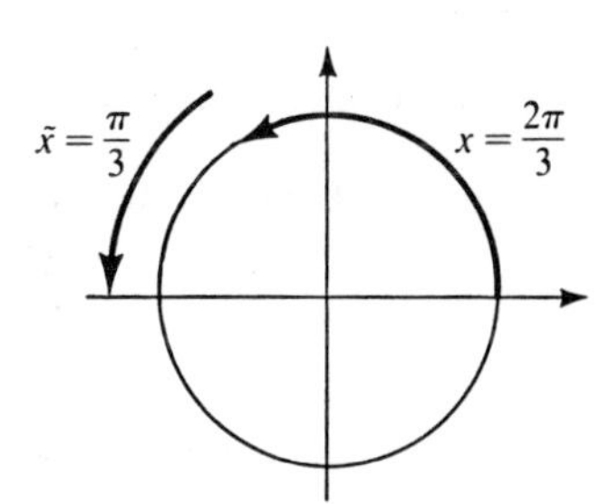

c.

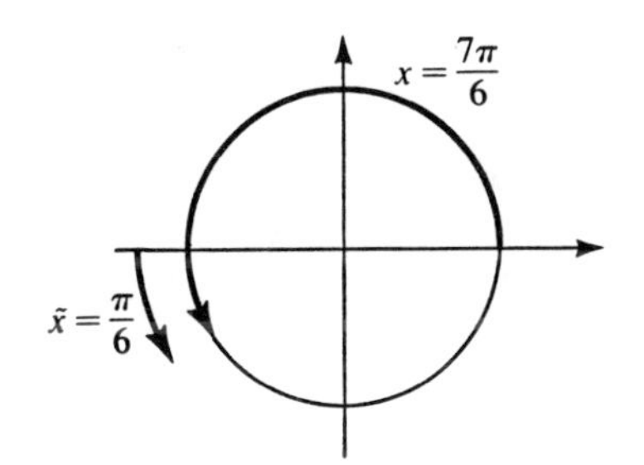

d.

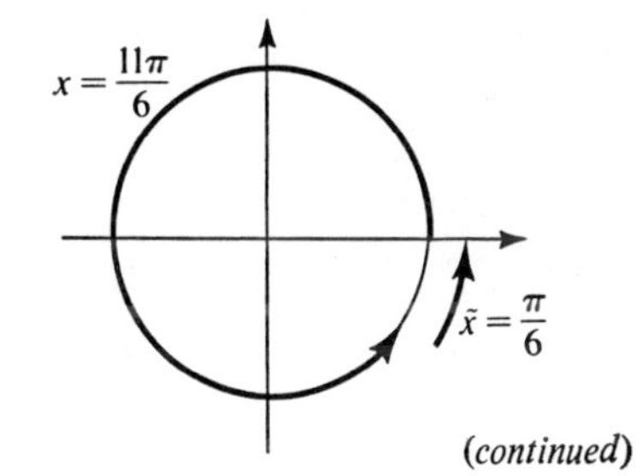

(continued)

e.

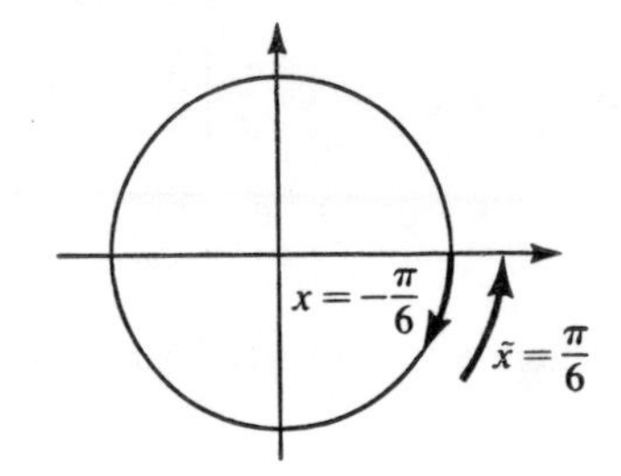

f.

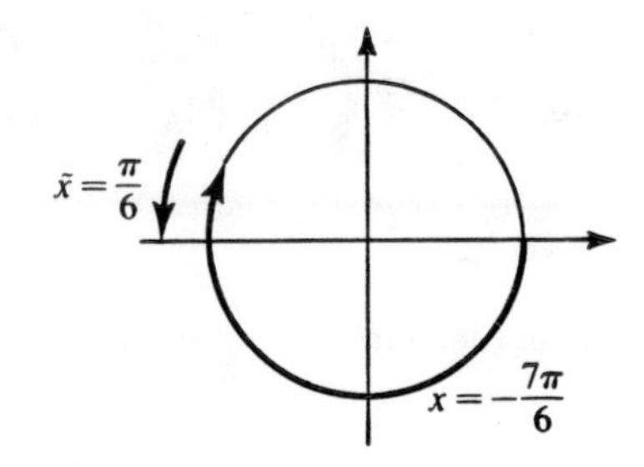

Properties

1. *Signs for circular function values:*

Quadrant	sin x csc x	cos x sec x	tan x cot x
I	+	+	+
II	+	−	−
III	−	−	+
IV	−	+	−

2. *Periods of the circular functions* ($a \in R$)
   a. *Sine, cosecant, cosine, and secant have period* 2π; *for these functions* 2π *is the smallest positive number* a *such that* **trig** x = **trig** $(x + a)$.
   b. *Tangent and cotangent have fundamental period* π; *for these functions* π *is the smallest positive number* a *such that* **trig** x = **trig** $(x + a)$.
3. *For each* x *such that* **trig** x *and* **trig** $\tilde{x}$ *are defined:*

$$\textbf{trig } x = \textbf{trig } \tilde{x} \quad \textit{if } \textbf{trig } x > 0 \quad \textit{or} \quad \textbf{trig } x = -\textbf{trig } \tilde{x} \quad \textit{if } \textbf{trig } x < 0$$

Exercises

Find each function value.

Example $\cos \frac{\pi}{3}$

Solution By definition, $\cos \frac{\pi}{3} = \cos \frac{\pi^R}{3}$. Since $\cos \frac{\pi^R}{3} = \frac{1}{2}$ (see Preliminary Concepts),

$$\cos \frac{\pi}{3} = \frac{1}{2}$$

1. $\tan\frac{\pi}{3}$ **2.** $\cos\frac{\pi}{4}$ **3.** $\sec\frac{\pi}{6}$

4. $\sin\frac{\pi}{2}$ **5.** $\cos 0$ **6.** $\tan\frac{3\pi}{2}$

Example $\sin\frac{7\pi}{6}$

Solution The use of a reference arc is helpful here. From the diagram at the right, $\tilde{x} = \frac{\pi}{6}$. By Property 3, $\sin\frac{7\pi}{6} = -\sin\frac{\pi}{6}$, where the negative sign is chosen because the arc with length $\frac{7\pi}{6}$ terminates in Quadrant III. Furthermore, $\sin\frac{\pi}{6} = \sin\frac{\pi^R}{6} = \frac{1}{2}$. Hence

$$\sin\frac{7\pi}{6} = -\sin\frac{\pi}{6} = -\frac{1}{2}$$

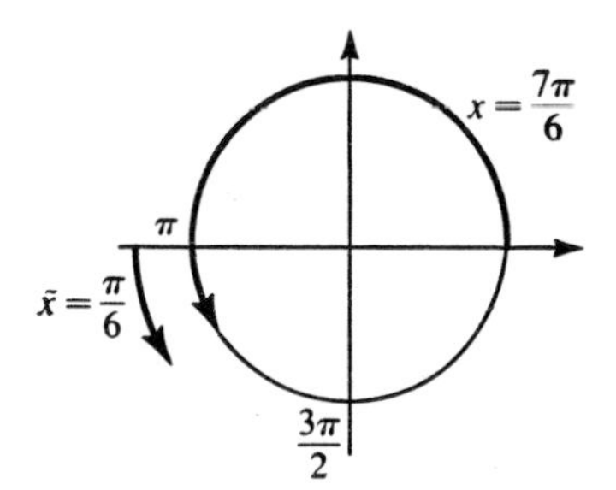

7. $\cos\frac{2\pi}{3}$ **8.** $\sin\frac{5\pi}{3}$ **9.** $\cot\frac{7\pi}{4}$

10. $\csc\frac{4\pi}{3}$ **11.** $\cos\frac{5\pi}{4}$ **12.** $\sin\frac{11\pi}{6}$

Find each function value. Use Table III *on page* 70 *as necessary.*

Examples **a.** sin 1.32 **b.** cos 2.04

Solutions **a.** From Table III, we have $\sin 1.32 \approx 0.9687$.

b. From the diagram at the right, $\tilde{x} \approx 1.10$. Then by Property 3, $\cos 2.04 = -\cos 1.10$, where the negative sign is chosen because the arc with length 2.04 terminates in Quadrant II. Furthermore from Table III,

$$\cos 1.10 \approx 0.4536$$

Hence $\cos 2.04 \approx -\cos 1.10 \approx -0.4536$.

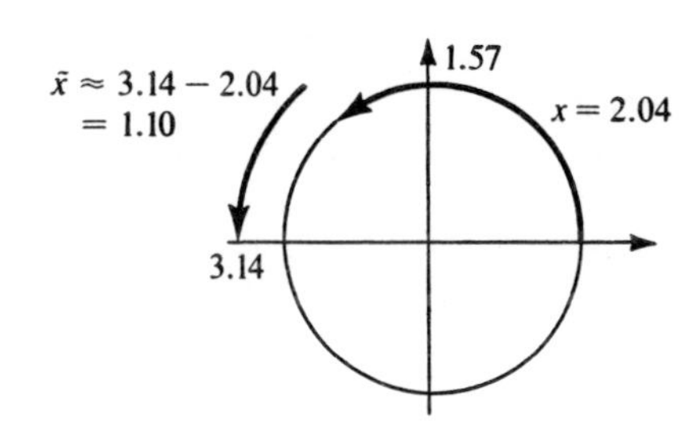

13. cos 0.43 **14.** sin 1.21 **15.** tan 1.93
16. tan 3.73 **17.** sin 5.83 **18.** cos 4.61
19. cos 1.92 **20.** sin 6.01 **21.** tan (−5.42)
22. tan (−2.45) **23.** cos (−3.92) **24.** sin (−2.91)

Example sin 10.67

Solution Since the sine function is periodic with period $2\pi \approx 6.28$,

$$\sin 10.67 \approx \sin (10.67 - 6.28) = \sin 4.39$$

Since the arc with length 4.39 terminates in Quadrant III, sin 4.39 is negative. Furthermore, from Table III, we have sin 1.25 = 0.9490. Hence

$$\begin{aligned} \sin 10.67 &= \sin 4.39 \\ &= -\sin 1.25 \\ &= -0.9490 \end{aligned}$$

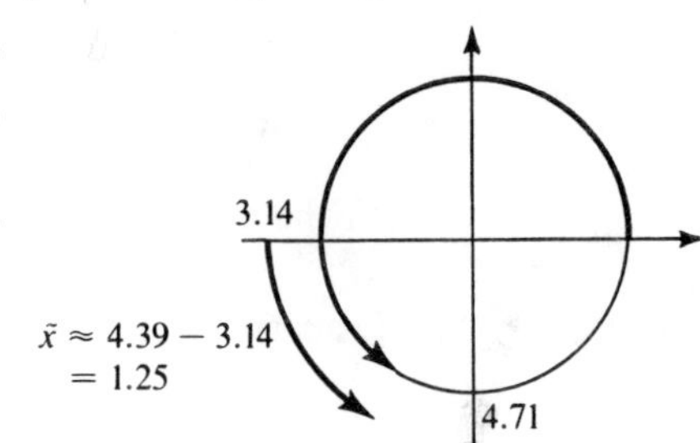

Note: Alternatively we could have shown $\tilde{x}$ in relation to $x = 10.67$.

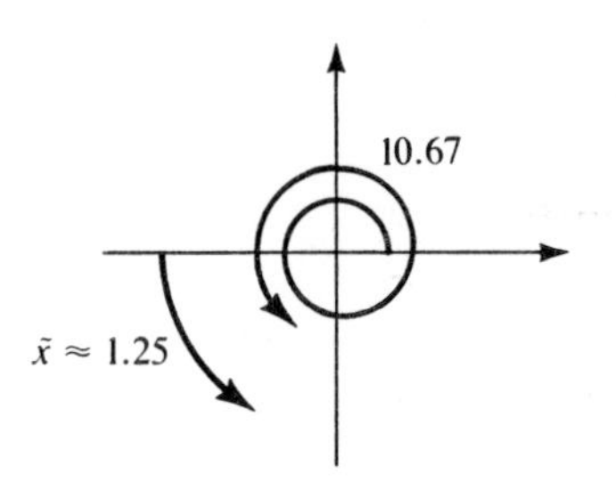

25. cos 7.21 **26.** tan 8.68 **27.** sin 11.30
28. cos 10.92 **29.** sin 15.60 **30.** tan 25.73

The equation $i = I_{max} \sin \omega t$ *is used to find the instantaneous current i (in amperes) at time t (in seconds).* I_{max} *and* ω *are constants for the particular circuit. For a given circuit it is found that* $I_{max} = 0.06$ *and* $\omega = 350$. *Find the current at each given time to the nearest thousandth.*

Example $t = 0.5$

Solution In $i = I_{max} \sin \omega t$, we substitute 0.06 for I_{max}, 350 for ω, and 0.5 for t and obtain

$$i = 0.06 \sin [350(0.5)] = 0.06 \sin 175$$

Since the sine function has period $2\pi \approx 6.28$ and $175 \div 6.28$ equals 27 with a remainder of 5.44,

$$\sin 175 \approx \sin [175 - 27(6.28)] = \sin 5.44$$

From the diagram at the right, $\tilde{x} \approx 0.84$. By Property 3,

$$\sin 5.44 = -\sin 0.84$$

where the negative sign is chosen because the arc with length 5.44 terminates in Quadrant IV. Furthermore, from Table III,

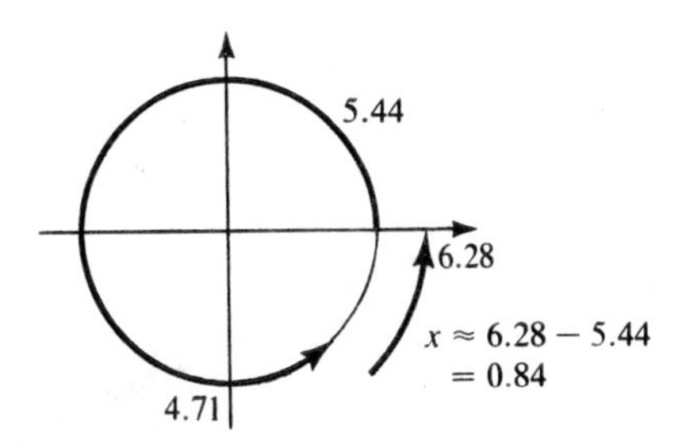

$$\sin 0.84 \approx 0.7446$$

Hence

$$\sin 5.44 \approx -\sin 0.84 \approx -0.7446$$

Thus we have

$$\begin{aligned} i = 0.06 \sin 175 &\approx 0.06 \sin [175 - 27(6.28)] \\ &= 0.06 \sin 5.44 \approx 0.06\,(-\sin 0.84) \\ &\approx 0.06(-0.7446) \approx -0.045 \end{aligned}$$

Hence, to the nearest thousandth, $i \approx -0.045$ ampere.

.06 sin 7

31. $t = 0.02$ **32.** $t = 0.06$ **33.** $t = 0.1$

34. $t = 0.6$ **35.** $t = 1.0$ **36.** $t = 2.5$

One end of a spring is attached to a ceiling and then displaced from its rest position. The equation $d = k \cos \omega t$ is used to find the displacement d of the spring from its position at rest to its position at time t. The constants k and ω depend upon the spring and the initial position of the spring at $t = 0$ and the initial displacement k. When t is measured in seconds, d is in centimeters. Let $\omega = 5$ for each spring in each of the following exercises. Use $\pi = 3.14$, and express answers to the nearest tenth.

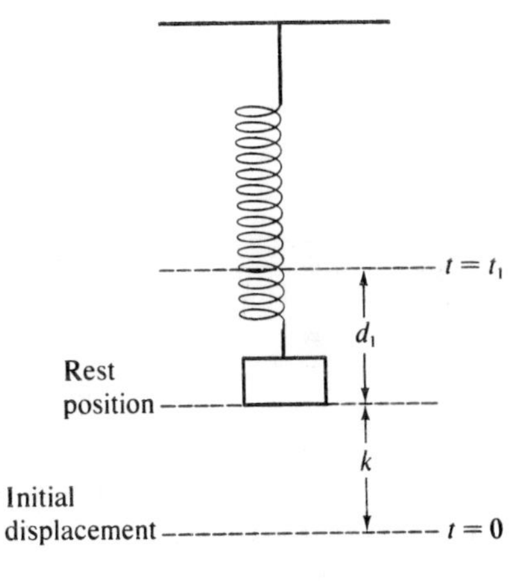

Example If the initial displacement k is 7, find d when $t = 4$.

Solution In the formula $d = k \cos \omega t$, we substitute 7 for k, 5 for ω, and 4 for t, and obtain

$$d = 7 \cos (5)(4) = 7 \cos 20$$

Since the cosine function has period $2\pi \approx 6.28$, and $20 \div 6.28$ equals 3 with a remainder of 1.16,

$$\cos 20 \approx \cos [20 - 3(6.28)] = \cos 1.16$$

From Table III, $\cos 1.16 \approx 0.3993$. Thus

$$d = 7 \cos 20 \approx 7 \cos 1.16$$
$$\approx 7(0.3993) \approx 2.8$$

Hence, to the nearest tenth, $d = 2.8$ centimeters.

37. If the initial displacement k is -8, find d when $t = 2$ and also when $t = 5$.

38. If the initial displacement k is 6, find d when $t = 3$ and also when $t = 5$.

39. If the initial displacement k is -6, find d when t has the values $0, 2\pi, 4\pi$.

40. If the initial displacement k is 10, find d when t has the values $0, 2\pi, 4\pi$.

41. If $d = 15$ when $t = 1$, find k, the initial displacement.

42. If $d = 10$ when $t = 3$, find k, the initial displacement.

The equation $d = A \sin 2\pi t$ describes the horizontal displacement d of the weight at the end of an oscillating pendulum. When A is in centimeters and t is in seconds, d is in centimeters. Use $\pi \approx 3.14$ and express answers to the nearest tenth.

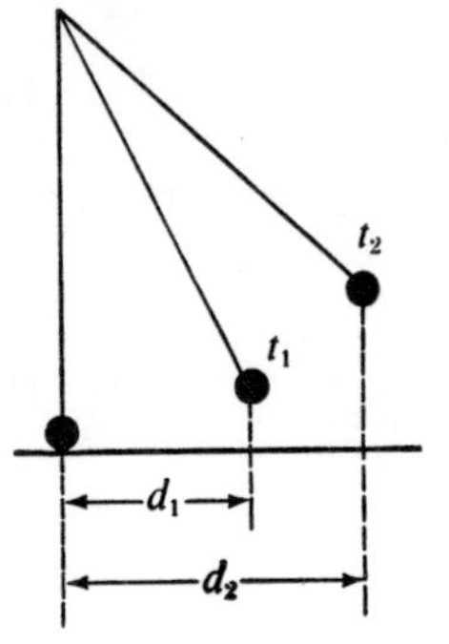

Example If $t = 1.4$ and $A = 26$, find d.

Solution In the formula $d = A \sin 2\pi t$, we substitute 1.4 for t and 26 for A and obtain

$$d = 26 \sin 2\pi(1.4) \approx 26 \sin 8.79$$

Since the sine function has period $2\pi \approx 6.28$,

$$\sin 8.79 \approx \sin (8.79 - 6.28) = \sin 2.51$$

From the diagram we note that $\bar{x} \approx 0.63$. Hence, by Property 3, $\sin 2.51 = \sin 0.63$, where the positive sign is chosen because the arc with length 2.51 terminates in Quadrant II. Furthermore, from Table III,

$$\sin 0.63 \approx 0.5891$$

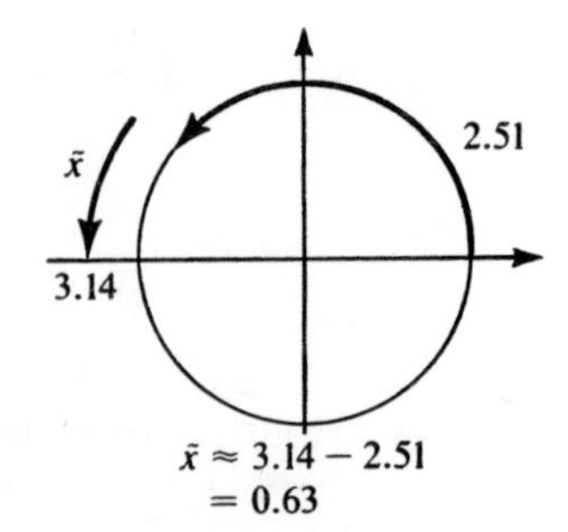

Hence $\sin 2.51 \approx \sin 0.63 \approx 0.5891$ and

$$d \approx 26 \sin 8.79 \approx 26 \sin 2.51$$
$$\approx 26 \sin 0.63 \approx 26(0.5891) \approx 15.3$$

Hence the displacement $d \approx 15.3$ centimeters.

43. If $t = 2.2$ and $A = 18$, find d.

44. If $t = 1.6$ and $A = 30$, find d.

45. If $d = 22$ when $t = \frac{1}{4}$, find A.

46. If $d = 5$ when $t = \frac{1}{6}$, find A.

1.2 GRAPHS OF SINE AND COSINE FUNCTIONS

Definitions

Sine wave — Graph of $y = A \sin B(x + C)$ or $y = A \cos B(x + C)$, where A, B, and C are constants

Amplitude — One-half of the absolute value of the difference of the maximum and minimum ordinates of a sine wave

Cycle — The portion of the graph of a periodic function over each period

Zeros of a function — The numbers in the domain of a function which are associated with 0 in its range

Example

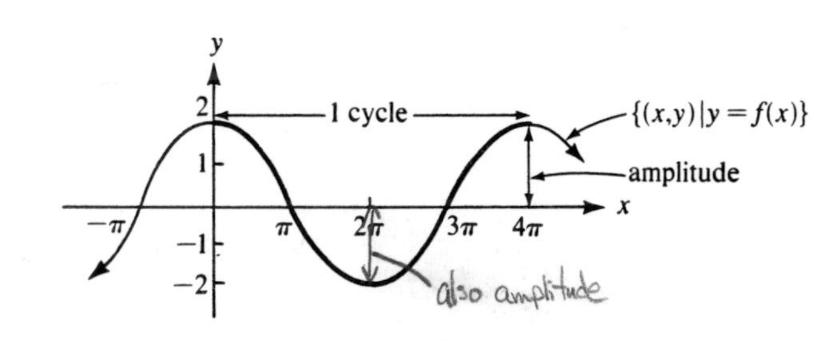

a. Since the maximum and minimum ordinates are 2 and −2, respectively, the amplitude is $\frac{1}{2}|2 - (-2)| = 2$.

b. From the graph, when $x \in \{\ldots, -\pi, \pi, 3\pi, \ldots,\}$, $f(x) = 0$. Hence, the zeros of $f(x)$ are $x = \pi + k \cdot 2\pi$, $k \in J$. Note that zeros name x intercepts.

Phase shift A number associated with a horizontal change (translation) in the graph of a periodic function

Example

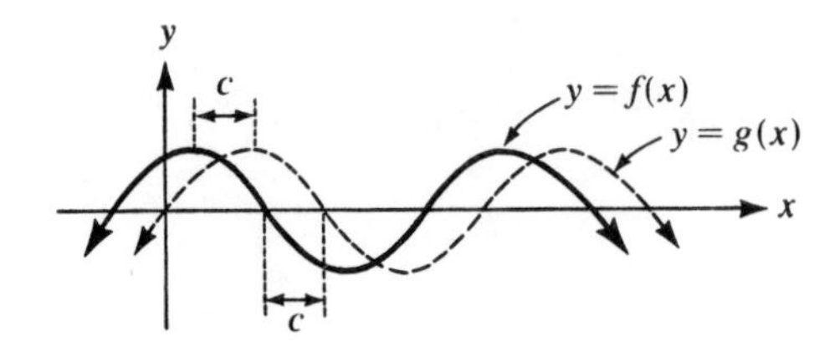

The phase shift of the graph of $y = f(x)$ relative to $y = g(x)$ is C.

Properties

1. *sine:*

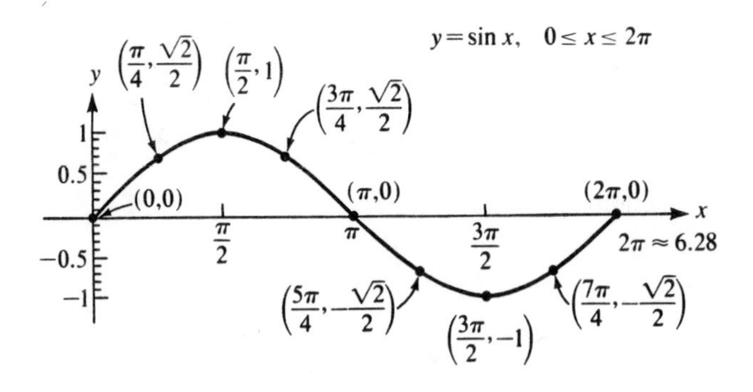

Note: $\frac{\sqrt{2}}{2} \approx 0.707$

$y = \sin x, \quad x \in R$

y
2
1
−1
−2
2
4
6
8
10
12
x
−π
−π/2
π/2
π
3π/2
2π
5π/2
3π
7π/2
4π

Amplitude: 1; Period: 2π

2. *cosine*:

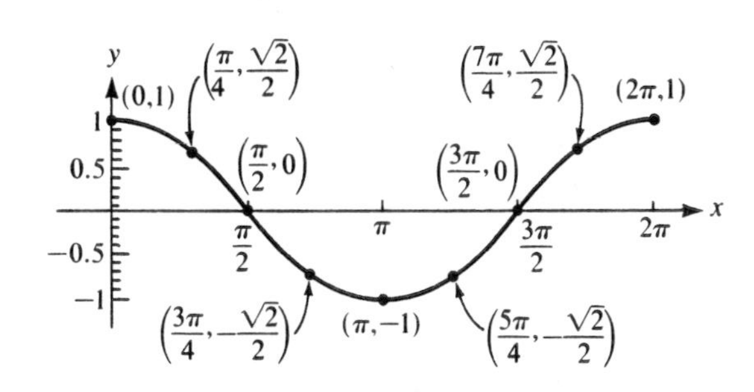

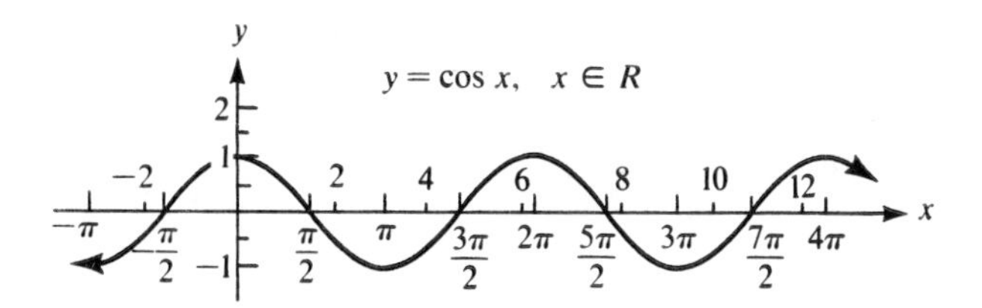

Amplitude A: 1; Period B: 2π

3. *If* $\boldsymbol{A,B,C \in R}$, *the graphs of the sine waves*

$$y = A \sin B(x + C) \quad \text{or} \quad y = A \cos B(x + C)$$

a. *have amplitude* $|A|$

b. *have a period* $p = \dfrac{2\pi}{|B|}$

c. *have phase shift* $|C|$; *if* $C > 0$, *the graphs are* $|C|$ *units to the left of the graphs of* $y = A \sin Bx$ *or* $y = A \cos Bx$; *if* $C < 0$, *the graphs are* $|C|$ *units to the right of the graphs of* $y = A \sin Bx$ *or* $y = A \cos Bx$.

Exercises

Sketch the graphs of the following equations over the interval $-2\pi \le x \le 2\pi$.

Example $y = 3 \sin x$

Solution It may be helpful first to sketch $y = \sin x$, $0 \le x \le 2\pi$, as a reference (dashed curve). Then, since the amplitude of $y = 3 \sin x$ is 3, we can sketch the desired graph on the same coordinate system over the interval $0 \le x \le 2\pi$ by making each ordinate 3 times the corresponding ordinate of the graph of $y = \sin x$. We can then extend this cycle to include the entire interval $-2\pi \le x \le 2\pi$, as shown in the figure. For $y = 3 \sin x$, the fundamental period p is the same as for $y = \sin x$, namely, $p = 2\pi$. A table showing values of y for selected values of x is helpful as a check.

x	$\sin x$	$3 \sin x$ or y
0	0	0
$\frac{\pi}{2}$	1	3
π	0	0
$\frac{3\pi}{2}$	−1	−3
2π	0	0

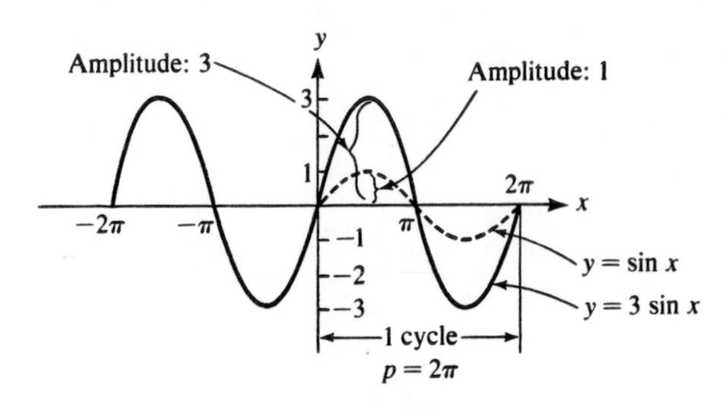

Note: Since the values of x chosen for tables are arbitrary, tables will not be shown in the Solution Key. However, the student should make them as a check.

1. $y = 2 \sin x$
2. $y = 3 \cos x$
3. $y = \frac{1}{2} \cos x$
4. $y = \frac{1}{3} \cos x$
5. $y = -4 \sin x$
6. $y = -\frac{1}{2} \cos x$

Example $y = \cos 2x$

Solution Let us first sketch a cycle of $y = \cos x$, $0 \le x \le 2\pi$, as a reference (dashed curve). Since

$$p = \frac{2\pi}{|B|} = \frac{2\pi}{2} = \pi$$

we next sketch a cycle of the graph of $y = \cos 2x$ over the interval $0 \le x \le \pi$ on the same coordinate system and extend the cycle obtained over the interval $-2\pi \le x \le 2\pi$. Again a table showing values of y for selected values of x may be helpful as a check.

x	$2x$	$\cos 2x$ or y
0	0	1
$\frac{\pi}{4}$	$\frac{\pi}{2}$	0
$\frac{\pi}{2}$	π	-1
$\frac{3\pi}{4}$	$\frac{3\pi}{2}$	0
π	2π	1

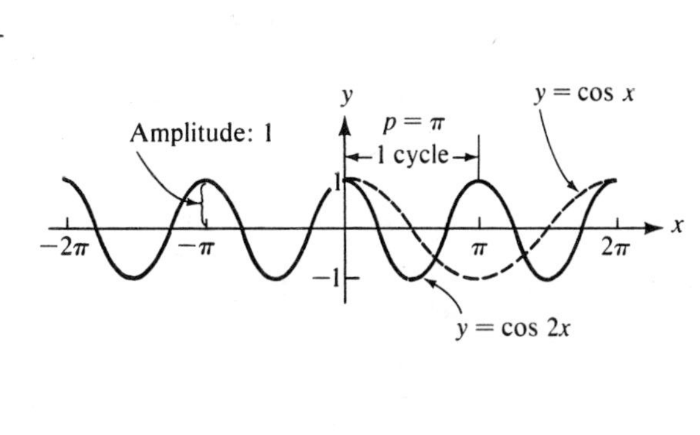

7. $y = \sin 2x$ **8.** $y = \cos 3x$ **9.** $y = \cos \frac{1}{3}x$ **10.** $y = \cos \frac{1}{2}x$

Example $y = -4 \sin \frac{1}{2}x$

Solution We first sketch the graph of $y = \sin x$ as a reference (dashed curve). Since $A = -4$, each ordinate of the graph of $y = -4 \sin (x/2)$ is the negative of the ordinate of the graph of $y = 4 \sin (x/2)$. Since

$$p = \frac{2\pi}{|B|} = \frac{2\pi}{1/2} = 4\pi$$

there is one cycle in the interval $0 \leq x \leq 4\pi$, or $-2\pi \leq x \leq 2\pi$.

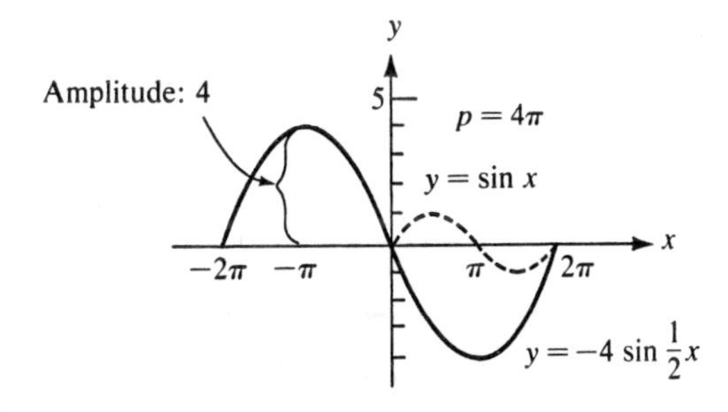

11. $y = -2 \cos x$ **12.** $y = -\cos 2x$

13. $y = -3 \sin 2x$ **14.** $y = -\frac{1}{2} \sin 3x$

Example $y = 3 \sin 2\left(x + \frac{\pi}{6}\right)$

Solution From Property 3, we note the following properties of this sine wave:

1. It has amplitude 3.
2. It has period $\frac{2\pi}{2} = \pi$.
3. It is $\frac{\pi}{6}$ units to the left of the graph of $y = 3 \sin 2x$.

First we sketch the graph of $y = 3 \sin 2x$ and then, with the above facts, we can sketch the graph shown.

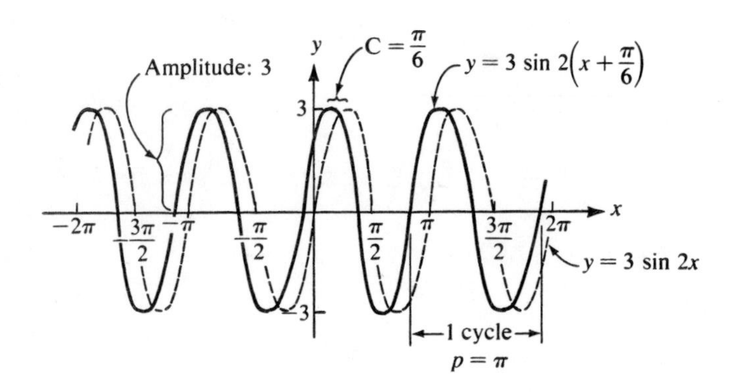

15. $y = \sin(x + \pi)$ **16.** $y = \cos\left(x - \frac{\pi}{2}\right)$ **17.** $y = 2\cos\left(x - \frac{\pi}{2}\right)$

18. $y = 3\sin\left(x + \frac{\pi}{6}\right)$ **19.** $y = 3\sin 2\left(x - \frac{\pi}{3}\right)$ **20.** $y = 2\cos 3\left(x + \frac{\pi}{4}\right)$

Example $y = \frac{1}{2}\sin\frac{\pi x}{2}$

Solution By Property 3, we note the following properties of the graph:

1. It is a sine wave.
2. It has amplitude $\frac{1}{2}$.
3. It has period $\frac{2\pi}{\pi/2} = 4$.

Since the period is 4, we use *integers* as elements of the domain. Scaling the x axis in *integral* units facilitates sketching the graph.

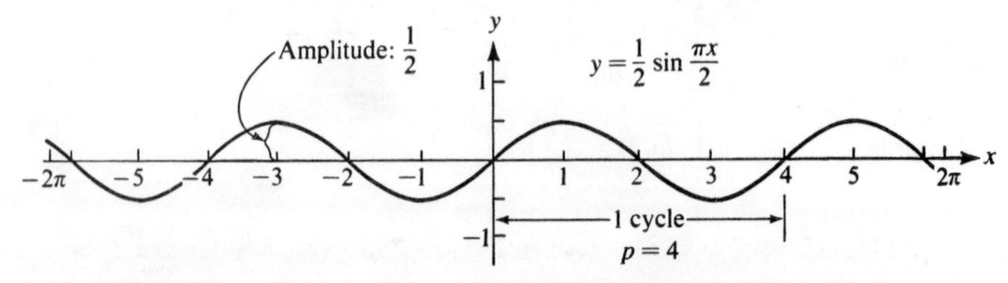

21. $y = 2 \sin \pi x$

22. $y = -3 \cos \frac{\pi}{2} x$

23. $y = -\frac{1}{2} \cos \frac{\pi}{3} x$

24. $y = \frac{1}{4} \sin \frac{\pi}{4} x$

From the respective graph, determine the zeros (over the specified interval) of the function defined by the equation in the given problem.

Example $y = \cos 2x$

Solution See graph of $y = \cos 2x$ on page 13. Zeros are the x intercepts

$$-\frac{7\pi}{4},\ -\frac{5\pi}{4},\ -\frac{3\pi}{4},\ -\frac{\pi}{4},\ \frac{\pi}{4},\ \frac{3\pi}{4},\ \frac{5\pi}{4},\ \frac{7\pi}{4}$$

25. Exercise 7 **26.** Exercise 8 **27.** Exercise 9

28. Exercise 10 **29.** Exercise 21 **30.** Exercise 22

1.3 GRAPHS OF THE TANGENT, COTANGENT, SECANT, AND COSECANT FUNCTIONS

Definition

Vertical asymptote A vertical line which the graph of $y = f(x)$ approaches

Examples a.

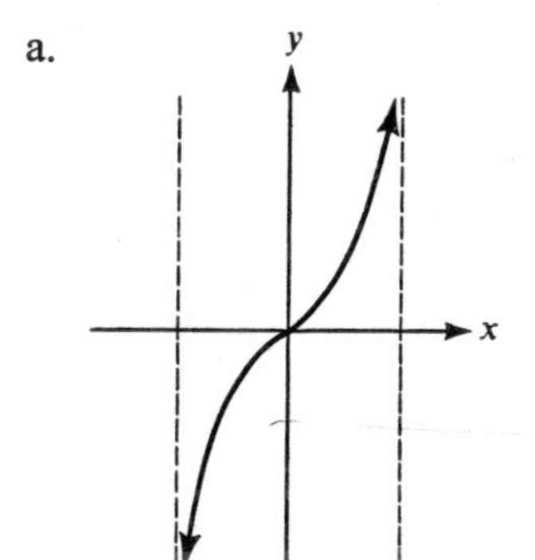

b.

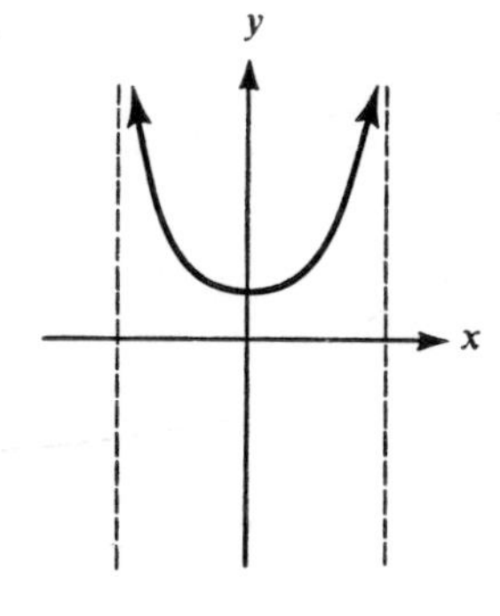

Dashed lines are vertical asymptotes of the respective graphs.

Properties ($A, B, C, a \in R$)

1. $\boldsymbol{x = a}$ *is the equation of a vertical asymptote of a trigonometric function defined by* $\boldsymbol{y = trig\ x}$ *for each* $\boldsymbol{a}$ *for which* $\boldsymbol{trig\ a}$ *is undefined.*

2. $\boldsymbol{y = \tan Bx}$ *and* $\boldsymbol{y = \cot Bx}$ *have period* $\boldsymbol{p = \dfrac{\pi}{|B|}}$;

   $\boldsymbol{y = \sec Bx}$ *and* $\boldsymbol{y = \csc Bx}$ *have period* $\boldsymbol{p = \dfrac{2\pi}{|B|}}$.

3. $\boldsymbol{y = A\ trig\ B(x + C)}$ *has phase shift* $\boldsymbol{|C|}$.

4. *The graph of the tangent function:*

$$\{(x,y)|y = \tan x, \quad 0 \le x \le \pi, \quad x \ne \frac{\pi}{2}\}$$

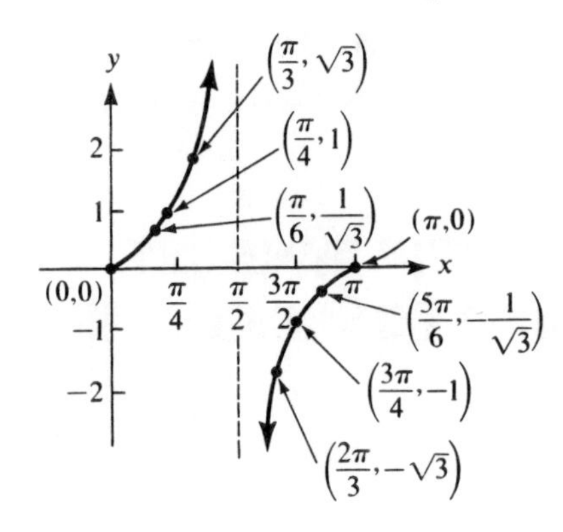

$$\{(x,y)|y = \tan x, \quad x \ne \frac{\pi}{2} + k\pi, \quad k \in J\}$$

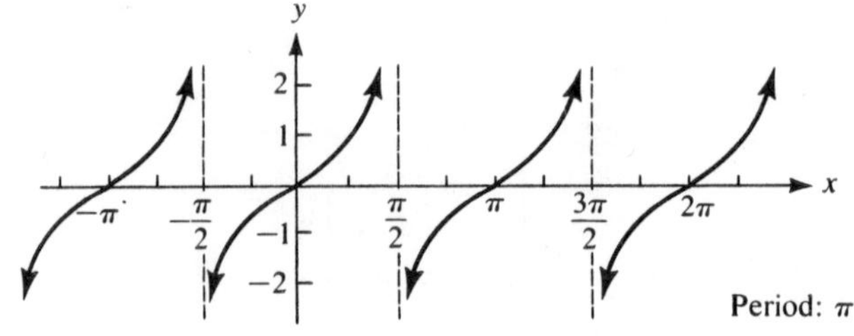

Period: π

Asymptotes: $x = \dfrac{\pi}{2} + k \cdot \pi, \quad k \in J$

Note: $\sqrt{3} \approx 1.7$ *and* $\dfrac{1}{\sqrt{3}} \approx 0.6$.

5. *The graphs of the reciprocal functions* $y = \cot x$ *and* $y = \tan x$:

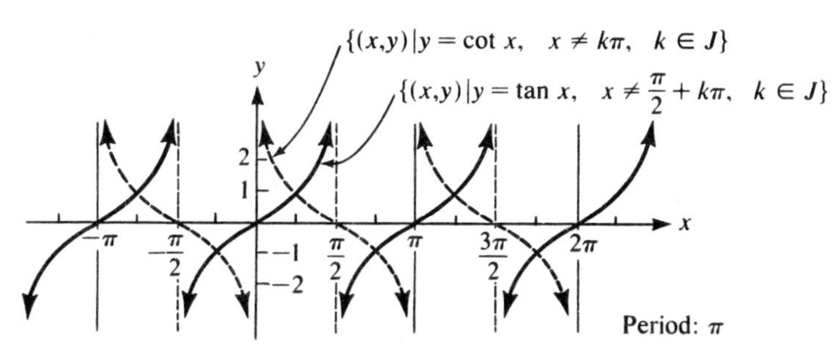

Note: For each x, $\cot x = \dfrac{1}{\tan x}$, $x \neq k \cdot \dfrac{\pi}{2}$, $k \in J$.

6. *The graphs of the reciprocal functions* $y = \sec x$ *and* $y = \cos x$:

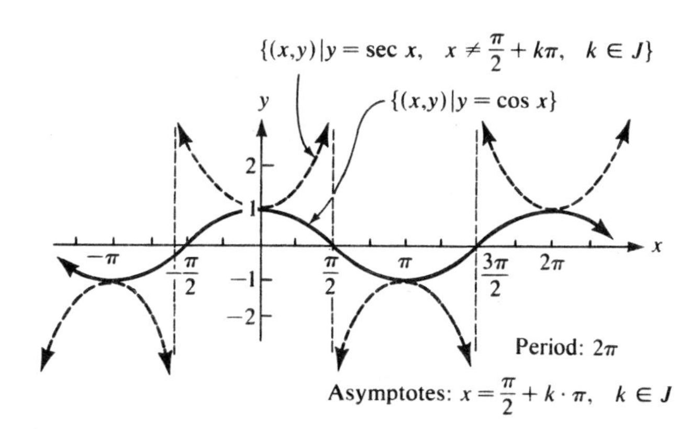

Note: For each x, $\sec x = \dfrac{1}{\cos x}$, $x \neq \dfrac{\pi}{2} + k \cdot \pi$, $k \in J$.

7. *The graphs of the reciprocal functions* $y = \csc x$ *and* $y = \sin x$:

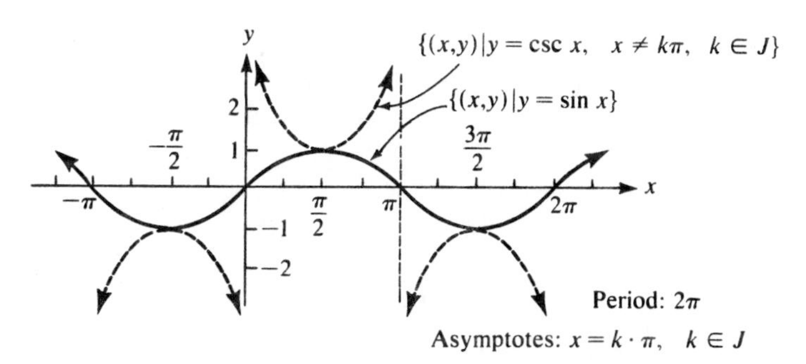

Note: For each x, $\csc x = \dfrac{1}{\sin x}$, $x \neq k \cdot \pi$, $k \in J$.

Exercises

Graph each of the following equations over the interval $-2\pi \le x \le 2\pi$.

Example $y = \csc \frac{1}{2} x$

Solution Since $\csc \frac{1}{2}x$ is undefined for $\frac{1}{2}x = k \cdot \pi$, $k \in J$, there are vertical asymptotes at $x = -2\pi, 0, 2\pi$. The period p is given by $p = \frac{2\pi}{|\frac{1}{2}|} = 4\pi$. Thus the graph of $y = \csc \frac{1}{2}x$ is similar to that of $y = \csc x$ (see Property 7), except each cycle is "spread out" over an interval of 4π units instead of 2π. The graph is sketched below. The table of values is a check.

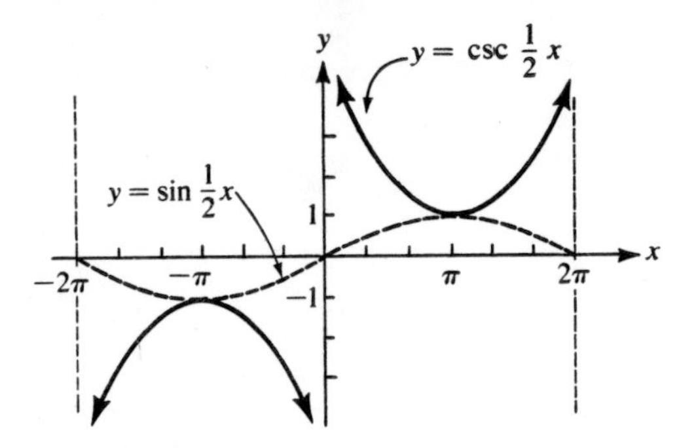

Note: For each x, $\csc \frac{1}{2}x = \frac{1}{\sin \frac{1}{2}x}$, $x \neq k \cdot 2\pi, k \in J$.

x	$\frac{1}{2}x$	y or $\csc \frac{1}{2}x$	(x, y)
-2π	$-\pi$	undef.	undef.
$\frac{-3\pi}{2}$	$\frac{-3\pi}{4}$	-1.4	$\left(-\frac{3\pi}{2}, -1.4\right)$
$-\pi$	$\frac{-\pi}{2}$	-1	$(-\pi, -1)$
$\frac{-\pi}{2}$	$\frac{-\pi}{4}$	-1.4	$\left(-\frac{\pi}{2}, -1.4\right)$
0	0	undef.	undef.
$\frac{\pi}{2}$	$\frac{\pi}{4}$	1.4	$\left(\frac{\pi}{2}, 1.4\right)$
π	$\frac{\pi}{2}$	1	$(\pi, 1)$
$\frac{3\pi}{2}$	$\frac{3\pi}{4}$	1.4	$\left(\frac{3\pi}{2}, 1.4\right)$

1. $y = \csc 2x$
2. $y = \sec 2x$
3. $y = \cot 2x$
4. $y = \tan 2x$
5. $y = \tan \frac{1}{2}x$
6. $y = \cot \frac{1}{2}x$
7. $y = 2 \sec \frac{1}{2}x$
8. $y = \frac{1}{2} \cot 2x$

Example $y = \tan\left(x - \frac{\pi}{4}\right)$

Solution Comparing $y = \tan\left(x - \frac{\pi}{4}\right)$ with $y = \tan x$ (see Property 3), we see that $\frac{\pi}{4}$ is the phase shift and the graph of $y = \tan\left(x - \frac{\pi}{4}\right)$ is $\frac{\pi}{4}$ units to the right of the graph of $y = \tan x$. First we sketch the graph of $y = \tan x$ and then shift it $\frac{\pi}{4}$ units to the right.

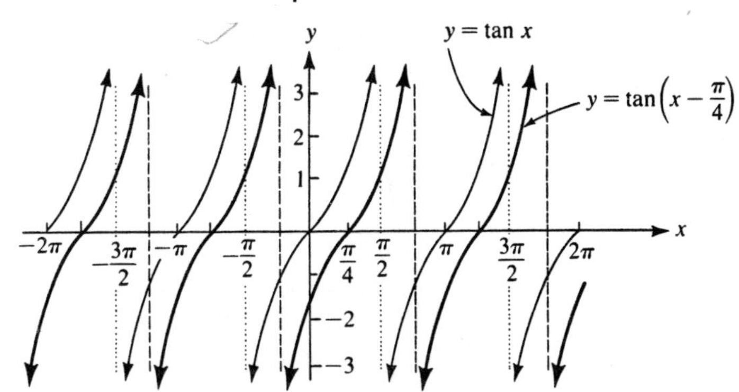

9. $y = \tan\left(x + \frac{\pi}{3}\right)$
10. $y = \cot\left(x - \frac{\pi}{3}\right)$
11. $y = \sec\left(x - \frac{\pi}{4}\right)$
12. $y = \csc\left(x + \frac{\pi}{4}\right)$

1.4 GRAPHICAL ADDITION

Exercises

Use the method shown in the following examples (called addition of ordinates) to sketch the graph of each equation over the interval $-2\pi \leq x \leq 2\pi$.

Example $y = \sin x + 2 \cos x$

Solution First, sketch the graphs of $y = \sin x$ and $y = 2 \cos x$ on the same coordinate system over $0 \le x \le 2\pi$. The ordinate of the graph of $y = \sin x + 2 \cos x$ at each point x on the x axis is the *algebraic* sum of the corresponding ordinates of $y = \sin x$ and $y = 2 \cos x$. For $x = \dfrac{\pi}{6}$,

$$\sin x = 0.5, \quad 2 \cos x = 2\left(\frac{\sqrt{3}}{2}\right) \approx 1.7, \quad \text{and} \quad \sin x + 2 \cos x \approx 2.2$$

The ordinate can be approximated graphically by "adding" the directed line segments from the x axis to the curve at $x = \dfrac{\pi}{6}$:

$$AB + AC = AB + BD = AD$$

If this is done for a few selected values of x, we obtain a good approximation for the curve. Then, duplicate this cycle of the curve over $-2\pi \le x < 0$.

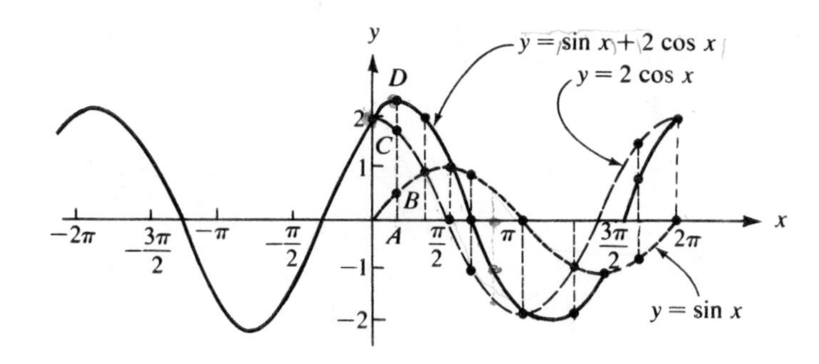

1. $y = \sin x + \cos x$

2. $y = 3 \sin x + \cos x$

3. $y = \sin 2x + \dfrac{1}{2} \cos x$

4. $y = \sin 3x + 2 \cos \dfrac{1}{2} x$

5. $y = \sin x - 2 \cos x$

6. $y = 3 \cos x - \sin 2x$

7. $y = 3 \sin x + 1$

8. $y = 4 \sin x - 3$

9. $y = x + \cos x$

10. $y = 2x - \cos x$

11. $y = x - \sin x$

12. $y = \dfrac{1}{3} x - 2 \cos x$

UNIT REVIEW

[**1.1**] *Find each function value.*

1. $\cos \frac{5\pi}{3}$ **2.** $\csc \frac{11\pi}{6}$ **3.** $\sin 1.86$

4. $\cot 5.64$ **5.** $\tan(-5.83)$ **6.** $\cos(-8.54)$

7. $\sec 10.38$ **8.** $\sin 15.48$

9. Use the equation $d = k \cos \omega t$ to find the instantaneous displacement (d) of an oscillating spring when $t = 2$ given that $\omega = 4$ and $k = -6$.

10. Use the equation $d = A \sin 2\pi t$ to find the instantaneous horizontal displacement (d) of the weight at the end of an oscillating pendulum when $t = 1.4$ given that $A = 20$.

[**1.2**] *Sketch the graphs of the following equations over the interval* $-2\pi \le x \le 2\pi$.

11. $y = \sin \frac{1}{2} x$ **12.** $y = \cos(x - \pi)$

13. $y = 2 \sin 2\left(x + \frac{\pi}{3}\right)$ **14.** $y = \frac{1}{2} \cos \frac{\pi}{4} x$

From the respective graph, determine the zeros (between -2π *and* 2π*) of the function defined by the equation.*

15. Exercise 13 above **16.** Exercise 14 above

[**1.3**] *Sketch the graph of each equation over the interval* $-2\pi \le x \le 2\pi$.

17. $y = \csc \frac{2}{3} x$ **18.** $y = \frac{1}{2} \sec 2x$

19. $y = \tan\left(x - \frac{\pi}{3}\right)$ **20.** $y = \cot\left(x + \frac{\pi}{4}\right)$

[**1.4**] *Sketch the graph of each equation over the interval* $-2\pi \le x \le 2\pi$.

21. $y = 2 \sin x - \cos 2x$ **22.** $y = x + \sin x$

UNIT 2

IDENTITIES AND CONDITIONAL EQUATIONS

The fact that the trigonometric functions are defined in terms of ratios which are determined by the coordinates of a point on the terminal side of an angle in standard position suggests that these functions are related in a number of ways. These relationships, in the form of equations, are called *identities.* We have considered several such relationships in Unit 1. In this unit we shall consider others. In addition, we shall consider methods of solving trigonometric *conditional equations.*

2.1 BASIC IDENTITIES

Definition

Identity An equation that is true for every replacement of any of the variables for which each member is defined

Properties*

For all angles x *(or real numbers* x*) for which each member is defined,*

1. $\sin^2 x + \cos^2 x = 1$
2. $\tan^2 x + 1 = \sec^2 x$
3. $\cot^2 x + 1 = \csc^2 x$
4. $\csc x = \dfrac{1}{\sin x}$

$\sin(-x) = -\sin x$ $\cos(-x) = \cos x$

* See Exercises 31 to 33.

5. $\sec x = \dfrac{1}{\cos x}$

6. $\cot x = \dfrac{1}{\tan x}$

7. $\tan x = \dfrac{\sin x}{\cos x}$

8. $\cot x = \dfrac{\cos x}{\sin x}$

Exercises

Verify that each of the following equations is an identity.

Example

$$\tan^2 x \cdot \frac{1}{1 + \tan^2 x} = \sin^2 x \qquad (1)$$

Solution

We shall simplify the left-hand member by writing each expression in terms of $\sin x$ and $\cos x$. Since $\tan x = \dfrac{\sin x}{\cos x}$, we have

$$\tan^2 x = \frac{\sin^2 x}{\cos^2 x}$$

and since $1 + \tan^2 x = \sec^2 x$ and $\sec^2 x = \dfrac{1}{\cos^2 x}$, we have

$$\tan^2 x \cdot \frac{1}{1 + \tan^2 x} = \sin^2 x \qquad (1)$$

$$\frac{\sin^2 x}{\cos^2 x} \cdot \frac{1}{\sec^2 x} = \sin^2 x$$

$$\frac{\sin^2 x}{\cos^2 x} \cdot \frac{1}{1/\cos^2 x} = \sin^2 x$$

$$\frac{\sin^2 x}{\cos^2 x} \cdot \cos^2 x = \sin^2 x$$

$$\sin^2 x = \sin^2 x$$

We have shown that the left-hand member of (1) is equivalent to the right-hand member. Hence Equation (1) is an identity.

Example

$$\frac{\sin x}{1 - \cos x} - \cot x = \frac{1}{\sin x} \qquad (2)$$

Solution We shall first rewrite the given equation in terms of $\sin x$ and $\cos x$ only. Substituting $\dfrac{\cos x}{\sin x}$ for $\cot x$, we obtain

$$\frac{\sin x}{1-\cos x} - \frac{\cos x}{\sin x} = \frac{1}{\sin x}$$

Writing the left-hand member as a single fraction,

$$\frac{\sin^2 x - (1-\cos x)\cos x}{(1-\cos x)\sin x} = \frac{1}{\sin x}$$

from which

$$\frac{\sin^2 x - \cos x + \cos^2 x}{(1-\cos x)\sin x} = \frac{1}{\sin x}$$

$$\frac{(\sin^2 x + \cos^2 x) - \cos x}{(1-\cos x)\sin x} = \frac{1}{\sin x}$$

$$\frac{1-\cos x}{(1-\cos x)\sin x} = \frac{1}{\sin x}$$

Since $1 - \cos x$ is restricted from 0 in (2) above, we can rewrite the left-hand member here and arrive at the equivalent equation

$$\frac{1}{\sin x} = \frac{1}{\sin x}$$

Hence, (2) is an identity.

1. $\cos x \tan x = \sin x$
2. $\sin x \sec x = \tan x$
3. $\sin x \cot x = \cos x$
4. $\tan x \csc x = \sec x$
5. $\sin^2 x \cot^2 x = \cos^2 x$
6. $\tan^2 x \cos^2 x = \sin^2 x$
7. $\cos^2 x(1 + \tan^2 x) = 1$
8. $(\csc^2 x - 1) = \cot^2 x$
9. $\sin x \sec x = \tan x$
10. $\cos x \tan x \csc x = 1$
11. $\sec x - \cos x = \sin x \tan x$
12. $\sec x - \sin x \tan x = \cos x$
13. $\sec^2 x - 1 = \dfrac{1}{\csc^2 x - 1}$
14. $\dfrac{\cos x - \sin x}{\cos x} = 1 - \tan x$
15. $\sec x \csc x - \cot x = \tan x$
16. $(1 - \cos^2 x)(1 + \cot^2 x) = 1$
17. $\dfrac{1 + \tan^2 x}{\tan^2 x} = \csc^2 x$
18. $\dfrac{\sin^2 x}{1 - \cos x} = \dfrac{1 + \sec x}{\sec x}$
19. $\tan^2 x - \sin^2 x = \sin^2 x \tan^2 x$
20. $\cot^2 x + \sec^2 x = \tan^2 x + \csc^2 x$
21. $\tan x + \sec x = \dfrac{1}{\sec x - \tan x}$
22. $\dfrac{\sec x + \csc x}{1 + \tan x} = \csc x$
23. $\dfrac{1 - \sec x}{\cos x - 1} = \sec x$
24. $\dfrac{1}{\csc x + \sin x + 2} = \dfrac{\sin x}{(1 + \sin x)^2}$

25. $\frac{\cos x}{\cot x - 6\cos x} = \frac{1}{\csc x - 6}$

26. $\sin x \tan^3 x = \frac{\tan^2 x - \sin^2 x}{\cos x}$

27. $\cos x = \frac{2\sin x + 3}{3\sec x + 2\tan x}$

28. $\frac{2\cos^2 x - 1}{3 - 2\sin^2 x} = \frac{2 - \sec^2 x}{3\sec^2 x - 2\tan^2 x}$

29. $\tan^4 x - \sec^4 x = 1 - 2\sec^2 x$

30. $\sin^4 x + 2\sin^2 x\cos^2 x = 1 - \cos^4 x$

Show that the property specified in each exercise is true.

31. Property 1. (*Hint:* From the figure, $\sin x = PB$ and $\cos x = OB$.)
32. Property 2. (*Hint:* Divide both members of $\cos^2 x + \sin^2 x = 1$ by $\cos^2 x$.)
33. Property 3. (*Hint:* Divide both members of $\cos^2 x + \sin^2 x = 1$ by $\sin^2 x$.)

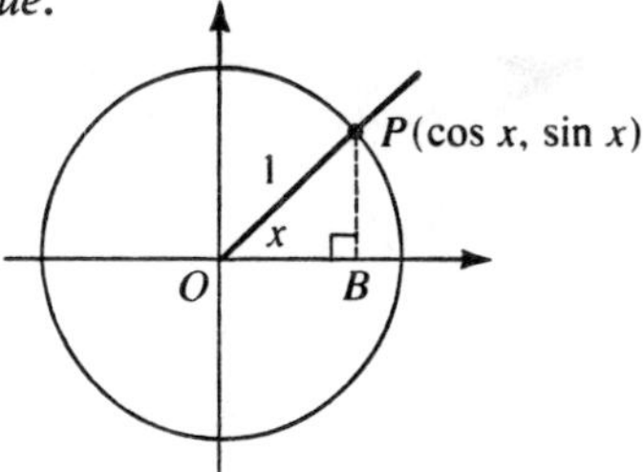

2.2 SUM FORMULAS, DIFFERENCE FORMULAS, AND SOME CONSEQUENCES

Definition

Reduction formula A trigonometric identity that expresses a relationship between a function value of an angle (or real number) which is not between 0 and $\pi/2$, and a function value of an angle (or real number) which is between 0 and $\pi/2$

Properties*

For all angles (or real numbers) x, x_1, x_2, for which each member is defined,

KNOW HOW TO USE THESE

1. $\cos(x_1 + x_2) = \cos x_1 \cos x_2 - \sin x_1 \sin x_2$
2. $\cos(x_1 - x_2) = \cos x_1 \cos x_2 + \sin x_1 \sin x_2$
3. $\sin(x_1 + x_2) = \sin x_1 \cos x_2 + \cos x_1 \sin x_2$
4. $\sin(x_1 - x_2) = \sin x_1 \cos x_2 - \cos x_1 \sin x_2$

* See Exercises 11 to 20 and 33 to 44.

5. $\cos 2x = \cos^2 x - \sin^2 x$
$= 2\cos^2 x - 1$
$= 1 - 2\sin^2 x$

6. $\sin 2x = 2 \sin x \cos x$

7. $\cos \frac{x}{2} = \pm\sqrt{\frac{1+\cos x}{2}}$

8. $\sin \frac{x}{2} = \pm\sqrt{\frac{1-\cos x}{2}}$

9. $\tan(x_1 + x_2) = \frac{\tan(x_1 + x_2)}{1 - \tan x_1 \tan x_2}$

10. $\tan(x_1 - x_2) = \frac{\tan x_1 - \tan x_2}{1 + \tan x_1 \tan x_2}$

11. $\tan 2x = \frac{2\tan x}{1 - \tan^2 x}$

12. $\tan \frac{x}{2} = \frac{1 - \cos x}{\sin x} = \frac{\sin x}{1 + \cos x}$

Reduction formulas (for all real numbers **x** *or all angles x with π replaced by* 180°):

13. $\sin(\pi - x) = \sin x$
14. $\cos(\pi - x) = -\cos x$
15. $\tan(\pi - x) = -\tan x$
16. $\sin(\pi + x) = -\sin x$
17. $\cos(\pi + x) = -\cos x$
18. $\tan(\pi + x) = \tan x$
19. $\sin(2\pi - x) = -\sin x$
20. $\cos(2\pi - x) = \cos x$
21. $\tan(2\pi - x) = -\tan x$
22. $\sin(-x) = -\sin x$
23. $\cos(-x) = \cos x$
24. $\tan(-x) = -\tan x$

Exercises

Example Given that $\sin x = \frac{4}{5}$, find $\sin 2x$, $\frac{\pi}{2} \le x \le \pi$.

Solution Since $\sin x = \frac{4}{5}$ and x terminates in Quadrant II,

$$\cos x = -\sqrt{1 - \sin^2 x} = -\sqrt{1 - \left(\frac{4}{5}\right)^2} = -\sqrt{\frac{9}{25}} = -\frac{3}{5}$$

From Property 6, we have

$$\sin 2x = 2 \sin x \cos x = 2\left(\frac{4}{5}\right)\left(-\frac{3}{5}\right) = -\frac{24}{25}$$

1. Given that $\cos x = -\frac{3}{5}$, find:

(*a*) $\cos 2x, \quad \frac{\pi}{2} \le x \le \pi$ (*b*) $\sin 2x, \quad \pi \le x \le \frac{3\pi}{2}$

(*c*) $\sin \frac{1}{2}x, \quad 3\pi \le x \le \frac{7\pi}{2}$ (*d*) $\cos \frac{1}{2}x, \quad \frac{\pi}{2} \le x \le \pi$

2. Given that $\sin x = \frac{\sqrt{5}}{3}$, find:

(*a*) $\cos 2x, \quad \frac{\pi}{2} \le x \le \pi$ (*b*) $\sin 2x, \quad \frac{\pi}{2} \le x \le \pi$

(*c*) $\sin \frac{1}{2}x, \quad 0 \le x \le \frac{\pi}{2}$ (*d*) $\cos \frac{1}{2}x, \quad 2\pi \le x \le \frac{5\pi}{2}$

Example Given that $\sin x = 0.2$, find an approximation for $\sin 2x$, $\frac{\pi}{2} \le x \le \pi$.

Solution Since $\sin x = 0.2$ and x terminates in Quadrant II,

$$\cos x = -\sqrt{1 - \sin^2 x} = -\sqrt{1 - 0.04} = -\sqrt{0.96} \approx -0.98$$

From Property 6,

$$\sin 2x = 2 \sin x \cos x \approx 2(0.2)(-0.98) \approx -0.39$$

3. Given that $\cos x = 0.7$, find an approximation for:

(*a*) $\sin 2x, \quad 0 \le x \le \frac{\pi}{2}$ (*b*) $\cos 2x, \quad \frac{3\pi}{2} \le x \le 2\pi$

(*c*) $\sin \frac{1}{2}x, \quad 0 \le x \le \frac{\pi}{2}$ (*d*) $\cos \frac{1}{2}x, \quad \frac{3\pi}{2} \le x \le 2\pi$

4. Given that $\sin x = -0.3$, find an approximation for:

(*a*) $\sin 2x, \quad \pi \le x \le \frac{3\pi}{2}$ (*b*) $\cos 2x, \quad \frac{3\pi}{2} \le x \le 2\pi$

(*c*) $\cos \frac{1}{2}x, \quad \pi \le x \le \frac{3\pi}{2}$ (*d*) $\sin \frac{1}{2}x, \quad \frac{3\pi}{2} \le x \le 2\pi$

Prove each identity.

Example $$\frac{\sin 2x}{1 + \cos 2x} = \tan x$$

Solution Replacing $\sin 2x$ and $\cos 2x$ in the given equation with $2 \sin x \cos x$ and $2\cos^2 x - 1$, respectively, we have

$$\frac{2 \sin x \cos x}{1 + (2\cos^2 x - 1)} = \tan x$$

$$\frac{2 \sin x \cos x}{2\cos^2 x} = \tan x$$

$$\frac{\sin x}{\cos x} = \tan x$$

$$\tan x = \tan x$$

5. $\sin 2x = \dfrac{2 \tan x}{1 + \tan^2 x}$

6. $\dfrac{2}{\sin 2x} = \tan x + \cot x$

7. $\cot x - \cot 2x = \csc 2x$

8. $\dfrac{2}{1 + \cos 2x} = \sec^2 x$

9. $\dfrac{1 + \cos 2x}{\sin 2x} = \cot x$

10. $\cos 2x = \dfrac{1 - \tan^2 x}{1 + \tan^2 x}$

Show that each reduction formula is true using Properties 1 *to* 4 *and* 9 *and* 10.

Example $\sin(\pi - x) = \sin x$

Solution Substitute π for x_1 and x for x_2 in Property 4.

$$\begin{aligned} \sin(\pi - x) &= \sin \pi \cos x - \cos \pi \sin x \\ &= 0 \cdot \cos x - (-1) \sin x \\ &= \sin x \end{aligned}$$

11. $\sin(\pi + x) = -\sin x$
12. $\sin(2\pi - x) = -\sin x$
13. $\cos(\pi - x) = -\cos x$
14. $\cos(\pi + x) = -\cos x$
15. $\cos(2\pi - x) = \cos x$
16. $\tan(\pi - x) = -\tan x$
17. $\tan(\pi + x) = \tan x$
18. $\tan(2\pi - x) = -\tan x$
19. $\sin(-x) = -\sin x$. [*Hint:* $\sin(-x) = \sin(0 - x)$.]
20. $\cos(-x) = \cos x$

Find each function value using Properties 13 *to* 24.

Examples **a.** $\sin \dfrac{5\pi}{6}$ **b.** $\tan 300°$

Solutions

a. $\sin\frac{5\pi}{6} = \sin\left(\pi - \frac{\pi}{6}\right)$

Using $\sin(\pi - x) = \sin x$ (Property 13),

$$\sin\frac{5\pi}{6} = \sin\frac{\pi}{6} = \frac{1}{2}$$

b. $\tan 300° = \tan(360 - 60)°$

Using $\tan(360° - x) = -\tan x$ (Property 21),

$$\tan 300° = -\tan 60° = -\sqrt{3}$$

Note that the same results could have been obtained by using the notion of a reference arc.

21. $\cos\frac{2\pi}{3}$	**22.** $\sin\frac{7\pi}{6}$	**23.** $\sin\frac{5\pi}{3}$
24. $\cos\frac{5\pi}{4}$	**25.** $\tan\frac{7\pi}{4}$	**26.** $\tan\frac{5\pi}{6}$
27. $\sin 225°$	**28.** $\cos 210°$	**29.** $\tan 120°$
30. $\tan 240°$	**31.** $\sin(-45°)$	**32.** $\cos(-60°)$

Show that the property specified in each exercise is true.

33. Property 1. [*Hint:* The figure at right shows selected points on the unit circle together with their coordinates (see Exercise 31 of Section 2.1). Since the length of arc from the point $P_2(1,0)$ to

$$P_5(\cos(x_1 + x_2), \sin(x_1 + x_2))$$

is $x_1 + x_2$, as is the length of arc from $P_1(\cos x_1, -\sin x_1)$ to $P_4(\cos x_2, \sin x_2)$, the chords P_2P_5 and P_1P_4 have equal length. Use the distance formula, which asserts that for the points $P_1(x_1,y_1)$ and $P_2(x_2,y_2)$, the length of segment P_1P_2 is $\sqrt{(x_2 - x_1)^2 + (y_2 - y_1)^2}$.]

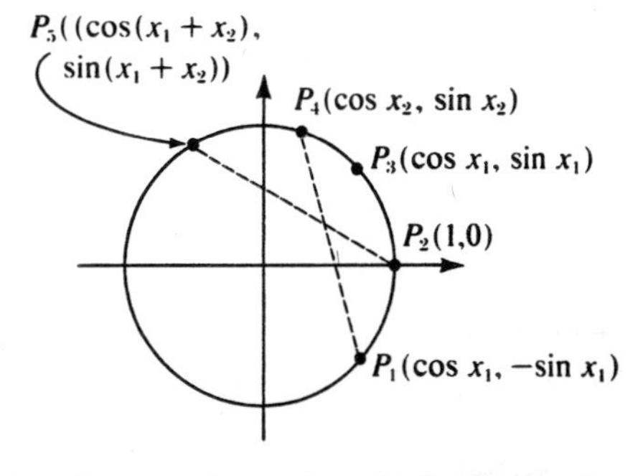

34. Property 2. (*Hint:* In Property 1, replace x_2 by $-x_2$.)

35. (*a*) For $x \in R$, $\cos\left(\frac{\pi}{2} - x\right) = \sin x$.

$\left(\textit{Hint:} \text{ In Property 2, replace } x_1 \text{ by } \frac{\pi}{2} \text{ and } x_2 \text{ by } x.\right)$

(*b*) For $x \in R$, $\sin\left(\frac{\pi}{2} - x\right) = \cos x$.

$\left(\textit{Hint:} \text{ In } (a) \text{ above, replace } x \text{ by } \frac{\pi}{2} - x.\right)$

(*c*) Property 3. $\Big($*Hint:* In the results of (*a*) above, replace x by $x_1 + x_2$ and write the resulting expression $\cos\left[\frac{\pi}{2} - (x_1 + x_2)\right]$ as $\cos\left[\left(\frac{\pi}{2} - x_1\right) - x_2\right]\Big)$.

36. Property 4. (*Hint:* In Property 3, replace x_2 by $-x_2$.)

37. Property 5. (*Hint:* In Property 1, replace x_1 and x_2 by x.)

38. Property 6. (*Hint:* In Property 3, replace x_1 and x_2 by x.)

39. Property 7. $\left(\textit{Hint: } \text{In } \cos 2x = 2\cos^2 x - 1, \text{ replace } x \text{ by } \frac{x}{2}.\right)$

40. Property 8. $\left(\textit{Hint: } \text{In } \cos 2x = 1 - 2\sin^2 x, \text{ replace } x \text{ by } \frac{x}{2}.\right)$

41. Property 9. $\left(\textit{Hint: } \tan(x_1 + x_2) = \frac{\sin(x_1 + x_2)}{\cos(x_1 + x_2)}.\right)$

42. Property 10. (*Hint:* In Property 9, replace x_2 by $-x_2$.)

43. Property 11. (*Hint:* In Property 9, replace x_1 and x_2 by x.)

44. Property 12. $\left(\textit{Hint: } \tan\frac{x}{2} = \frac{\sin x/2}{\cos x/2}.\right)$

2.3 CONDITIONAL EQUATIONS

Definition

Conditional equations	Equations that are not identities

Exercises

In all the following exercises, use Table II *or* III *as necessary, and give results to the nearest reading in the table.*

Find the solution set of each equation in (*a*) *radians and* (*b*) *degrees.*

Example $\tan x = -1$

Solution From memory or the table in Preliminary Concepts, we obtain

(*a*) $\left\{x \mid x = \frac{3\pi^R}{4} + k\pi^R, \quad k \in J\right\}$

(*b*) $\{x \mid x = 135° + k \cdot 180°, \quad k \in J\}$

1. $\cos x = \frac{1}{2}$ **2.** $\sin x = \frac{\sqrt{2}}{2}$ **3.** $\tan x = \sqrt{3}$

4. $\cot x = -1$ **5.** $\sec x - \sqrt{2} = 0$ **6.** $\csc x + 1 = 0$

7. $4 \sin x - 1 = 0$ **8.** $2 \tan x + 3 = 0$ **9.** $3 \cot x - 1 = 0$

10. $3 \cos x + 1 = 0$ **11.** $2 \sec x - 5 = 0$ **12.** $3 \csc x + 8 = 0$

Find the solution set of each equation in radians.

Example $\sqrt{3} \sin x + \cos x = 0$

Solution We shall first express the left-hand member in terms of *one* function value. Noting that $\cos x \neq 0$ in any solution, we can divide each member by $\cos x$ to produce

$$\sqrt{3} \frac{\sin x}{\cos x} + 1 = 0$$

from which

$$\sqrt{3} \tan x + 1 = 0$$

$$\tan x = -\frac{1}{\sqrt{3}}$$

Therefore, the solution set is $\left\{x \mid x = \frac{5\pi^R}{6} + k\pi^R, \quad k \in J\right\}$.

13. $\sin x - \sqrt{3} \cos x = 0$ **14.** $3 \sin x + \cos x = 0$

15. $\tan^2 x - 1 = 0$ **16.** $2 \sin^2 x - 1 = 0$

17. $\sin^2 x - \cos^2 x = 1$ **18.** $\sin x + \cos x \tan x = 3$

Find the solution set of each equation for $0° \leq \theta < 360°$.

Example $2 \cos^2 \theta \tan \theta - \tan \theta = 0$

Solution Factoring $\tan \theta$ from the terms in the left-hand member, we have

$$\tan \theta(2 \cos^2 \theta - 1) = 0$$

from which either

$$\tan \theta = 0 \quad \text{or} \quad 2 \cos^2 \theta - 1 = 0$$

If $\tan \theta = 0$, then either $\theta = 0°$ or $\theta = 180°$. If $2 \cos^2 \theta - 1 = 0$, then $\cos^2 \theta = 1/2$, $\cos \theta = \pm 1/\sqrt{2}$, and θ is an odd multiple of 45°. Therefore, the solution set over $0° \leq \theta < 360°$ is

$$\{0°, 45°, 135°, 180°, 225°, 315°\}$$

19. $(2 \sin \theta - 1)(2 \sin^2 \theta - 1) = 0$

20. $(\tan \theta - 1)(2 \cos \theta + 1) = 0$

21. $2 \sin \theta \cos \theta + \sin \theta = 0$

22. $\tan \theta \sin \theta - \tan \theta = 0$

Find the solution set of each equation over the interval $0^R \leq \alpha < 2\pi^R$.

Example $\tan^2 \alpha + \sec \alpha - 1 = 0$

Solution Since $\tan^2 \alpha = \sec^2 \alpha - 1$, the given equation can be written as

$$\sec^2 \alpha - 1 + \sec \alpha - 1 = 0$$
$$\sec^2 \alpha + \sec \alpha - 2 = 0$$
$$(\sec \alpha + 2)(\sec \alpha - 1) = 0$$

Now observe that the solution set is

$$\{\alpha \mid \sec \alpha + 2 = 0\} \cup \{\alpha \mid \sec \alpha - 1 = 0\}^*$$

over the interval $0^R \leq \alpha < 2\pi^R$. If $\sec \alpha + 2 = 0$, then $\sec \alpha = -2$, and if $\sec \alpha - 1 = 0$, then $\sec \alpha = 1$. The required solution set is

$$\left\{\frac{2\pi^R}{3}, \frac{4\pi^R}{3}\right\} \cup \{0^R\} = \left\{0^R, \frac{2\pi^R}{3}, \frac{4\pi^R}{3}\right\}$$

23. $\tan^2 \alpha - 2 \tan \alpha + 1 = 0$

24. $4 \sin^2 \alpha - 4 \sin \alpha + 1 = 0$

25. $\cos^2 \alpha + \cos \alpha = 2$

26. $\cot^2 \alpha = 5 \cot \alpha - 4$

27. $\sec^2 \alpha + 3 \tan \alpha - 11 = 0$

28. $\tan^2 \alpha + 4 = 2 \sec^2 \alpha$

29. $\sin^2 \alpha + \sin \alpha - 1 = 0$

30. $\tan^2 \alpha = \tan \alpha + 3$

(*Hint:* In Exercises 29 and 30 use the quadratic formula.)

Find the solution set of each equation in radians.

Example $\cos 2x = \sin x$

Solution Since $\cos 2x = 1 - 2 \sin^2 x$, the equation $\cos 2x = \sin x$ can be written equivalently as

$$1 - 2 \sin^2 x = \sin x$$
$$2 \sin^2 x + \sin x - 1 = 0$$
$$(2 \sin x - 1)(\sin x + 1) = 0$$

from which

$$\sin x = \frac{1}{2} \quad \text{or} \quad \sin x = -1$$

* "$\cup$" is read "union" and $\{\alpha \mid \sec \alpha + 2 = 0\} \cup \{\alpha \mid \sec \alpha - 1 = 0\}$ means that α is a member of $\{\alpha \mid \sec \alpha + 2 = 0 \text{ or } \sec \alpha - 1 = 0\}$,

Then as solution set we have

$$\left\{x \mid x = \frac{\pi^R}{6} + k \cdot 2\pi^R\right\} \cup \left\{x \mid x = \frac{5\pi^R}{6} + k \cdot 2\pi^R\right\}$$
$$\cup \left\{x \mid x = \frac{3\pi^R}{2} + k \cdot 2\pi^R\right\}, \qquad k \in J$$

31. $\sin 2x - \cos x = 0$

32. $\cos 2x = \cos^2 x - 1$

33. $\cos 2x = \cos x - 1$

34. $\sin x = \sin 2x$

In Exercises 35 *to* 40, *find the solution set of each equation over R.*

Example $\sqrt{2} \cos 3x = 1$

Solution The equation can be written equivalently as $\cos 3x = 1/\sqrt{2}$. Therefore,

$$3x = \frac{\pi}{4} + k \cdot 2\pi \quad \text{or} \quad 3x = \frac{7\pi}{4} + k \cdot 2\pi$$

from which, by dividing each member by 3, we obtain

$$x = \frac{\pi}{12} + k \cdot \frac{2\pi}{3} \quad \text{or} \quad x = \frac{7\pi}{12} + k \cdot \frac{2\pi}{3}, \quad k \in J$$

The solution set is $\left\{x \mid x = \frac{\pi}{12} + k \cdot \frac{2\pi}{3}\right\} \cup \left\{x \mid x = \frac{7\pi}{12} + k \cdot \frac{2\pi}{3}\right\}, \quad k \in J.$

35. $\cos 2x = \frac{\sqrt{2}}{2}$

36. $\tan 2x = \sqrt{3}$

37. $\cos 2x \sin x + \sin x = 0$

38. $\sin 2x \cos x - \sin x = 0$

39. $\sin 4x - 2 \sin 2x = 0$
[*Hint:* $\sin 4x = \sin 2(2x)$.]

40. $\sin 3x + 4 \sin^2 x = 0$
[*Hint:* $\sin 3x = \sin (x + 2x)$.]

2.4 INVERSE FUNCTIONS

Definitions and Notation

Inverse of a function, f

f^{-1}

The relation obtained by interchanging the components of each ordered pair of the function; the relation defined by the equation obtained by interchanging the variables in the equation that defines the function

Inverse trigonometric functions The six inverse trigonometric relations in which each range has been restricted so that each relation is a function

Arcsine

$$\text{Arcsine} = \left\{(x,y)\,|\,x = \sin y,\quad -\frac{\pi}{2} \le y \le \frac{\pi}{2}\right\}$$
$$= \{(x,y)\,|\,y = \text{Arcsin } x\}^*$$

Arccosine

$$\text{Arccosine} = \{(x,y)\,|\,x = \cos y,\quad 0 \le y \le \pi\}$$
$$= \{(x,y)\,|\,y = \text{Arccos } x\}$$

Arctangent†

$$\text{Arctangent} = \left\{(x,y)\,|\,x = \tan y,\quad -\frac{\pi}{2} < y < \frac{\pi}{2}\right\}$$
$$= \{(x,y)\,|\,y = \text{Arctan } x\}$$

Sin^{-1} x, Cos^{-1} x, etc. Arcsin x, Arccos x, etc.

Properties

1. *The graph of the inverse sine relation is the reflection of the graph of the sine function about the graph of* $y = x$.

   A similar relationship holds between the graphs of each of the other trigonometric functions and their respective inverse relations.

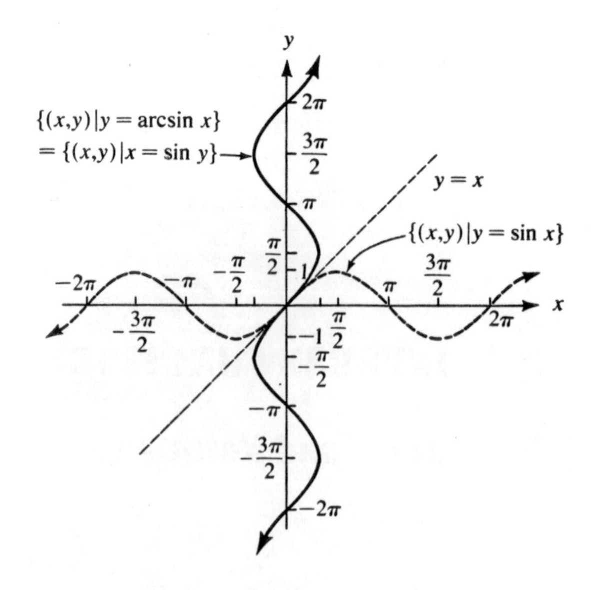

* If y is an angle, then $y = \text{Arcsin } x$ can be read "y is the angle whose sine is x." A similar comment holds for the other inverse trigonometric functions.

† Arccotangent, Arcsecant, and Arccosecant functions will not be defined since they are rarely used.

2. *Graphs of the inverse trigonometric functions:*

a.

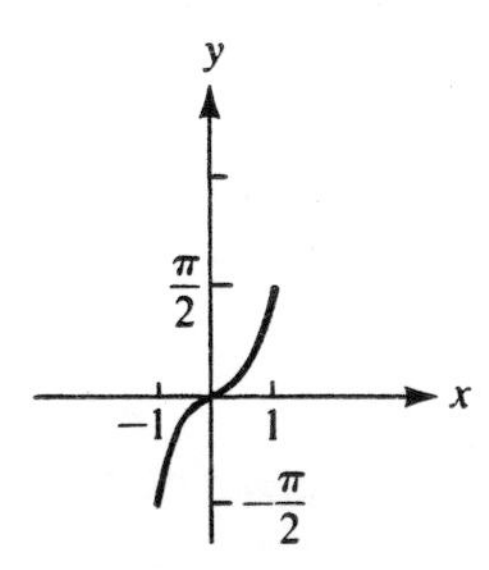

$$\left.\begin{array}{l} x = \sin y \\ y = \sin^{-1} x \end{array}\right\} -\frac{\pi}{2} \le y \le \frac{\pi}{2}$$

b.

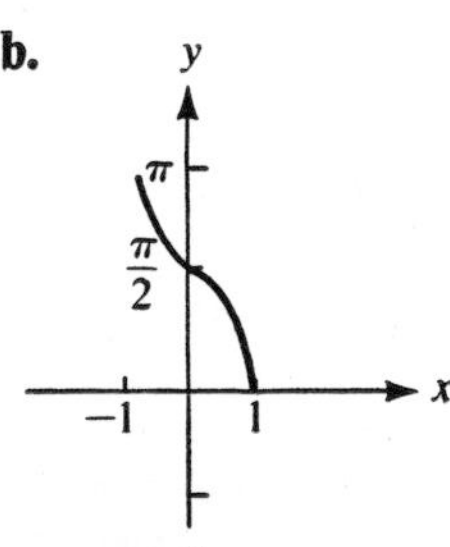

$$\left.\begin{array}{l} x = \cos y \\ y = \cos^{-1} x \end{array}\right\} 0 \le y \le \pi$$

c.

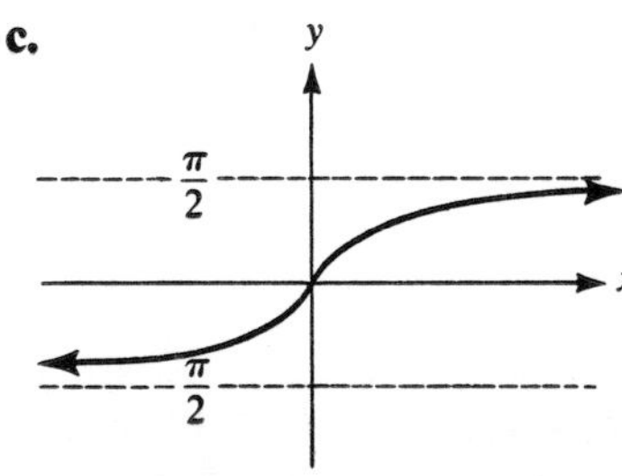

$$\left.\begin{array}{l} x = \tan y \\ y = \tan^{-1} x \end{array}\right\} -\frac{\pi}{2} < y < \frac{\pi}{2}$$

Exercises

Find the value of each expression if it exists. Express results in terms of rational multiples of π. Use the table in Preliminary Concepts as needed.

Examples

a. $\text{Arccos}\,\frac{1}{2}$ **b.** $\text{Tan}^{-1}(-1)$

Solutions

a. It may be helpful first to let $y = \text{Arccos}\,\frac{1}{2}$ and then to rewrite the equation equivalently as

$$\cos y = \frac{1}{2}, \quad 0 \le y \le \pi$$

From the table, or from memory, we have $y = \pi/3$. Hence,

$$\text{Arccos}\,\frac{1}{2} = \frac{\pi}{3}$$

b. Let $y = \text{Tan}^{-1}(-1)$. Hence,

$$\tan y = -1, \quad -\frac{\pi}{2} < y < \frac{\pi}{2}$$

from which $y = -\pi/4$. Hence

$$\text{Tan}^{-1}(-1) = -\frac{\pi}{4}$$

1. $\text{Arcsin}\,\frac{1}{2}$ **2.** $\text{Actan}\,\sqrt{3}$ **3.** $\text{Tan}^{-1}(-\sqrt{3})$ **4.** $\text{Cos}^{-1}\,\frac{1}{\sqrt{2}}$

5. $\text{Cos}^{-1}\,2$ **6.** $\text{Sin}^{-1}(-1)$ **7.** $\text{Arctan}\left(-\frac{1}{\sqrt{3}}\right)$ **8.** $\text{Arcsin}\,\frac{\sqrt{3}}{2}$

Use Table III on page 70 as necessary to find the value of each expression. Express results correct to the nearest entry in the table.

Example Arctan 0.2236

Solution Let $y = \text{Arctan } 0.2236$ and then rewrite the equation equivalently as

$$\tan y = 0.2236, \quad -\frac{\pi}{2} < y < \frac{\pi}{2}$$

From Table III, we find that $y \approx 0.22$. Since $-\pi/2 < 0.22 < \pi/2$,

$$\text{Arctan } 0.2236 \approx 0.22$$

9. Arctan 0.1003
10. Arccos (-0.3902)
11. $\text{Sin}^{-1}\, 0.3802$
12. $\text{Cos}^{-1}\, 0.6675$
13. Arcsin (-0.8624)
14. $\text{Tan}^{-1}\, 3.467$

Find the value for each expression if it exists.

Examples

a. $\text{Cos}^{-1}(\tan \pi)$

b. $\cos\left(\text{Sin}^{-1}\frac{1}{2}\right)$

Solutions

a. Since $\tan \pi = 0$,

$$\text{Cos}^{-1}(\tan \pi) = \text{Cos}^{-1}\, 0 = \frac{\pi}{2}$$

b. $\text{Sin}^{-1}\frac{1}{2} = \frac{\pi}{6}$. Hence,

$$\cos\left(\text{Sin}^{-1}\frac{1}{2}\right) = \cos\frac{\pi}{6} = \frac{\sqrt{3}}{2}$$

15. $\text{Sin}^{-1}\left(\cos\frac{\pi}{4}\right)$
16. $\text{Cos}^{-1}\left(\sin\frac{\pi}{2}\right)$
17. $\text{Tan}^{-1}\left(\tan\frac{\pi}{3}\right)$
18. $\text{Sin}^{-1}\left(\sin\frac{3\pi}{2}\right)$
19. $\sin\left(\text{Arccos}\frac{1}{2}\right)$
20. $\tan\left(\text{Arcsin}\frac{\sqrt{3}}{2}\right)$
21. Arccos (sin (Arctan (-1)))
22. $\sin(\text{Cos}^{-1}(\tan 0))$

Solve each equation over the intervals indicated in the definitions on page 34 specifying the solution using inverse notation. Do not use tables.

Example $3 \sin 2\theta = 1$

Solution Multiplying each member by $\frac{1}{3}$ yields $\sin 2\theta = \frac{1}{3}$. Hence

$$2\theta = \text{Sin}^{-1}\frac{1}{3}, \quad -\frac{\pi}{2} \le 2\theta \le \frac{\pi}{2}$$

Then, multiplying each member by $\frac{1}{2}$, we have

$$\theta = \frac{1}{2}\text{Sin}^{-1}\frac{1}{3}, \quad -\frac{\pi}{4} \le \theta \le \frac{\pi}{4}$$

23. $3 \sin \theta = 2$

24. $2 \tan \theta = 3$

25. $\cos 3\theta = \frac{1}{4}$

26. $\sin 2\theta = \frac{1}{3}$

27. $3 \tan 2\theta = 4$

28. $3 \cos 4\theta = 2$

29. $\frac{1}{2} \sin \frac{\theta}{3} = 0.28$

30. $\frac{1}{3} \cos \frac{\theta}{2} = 2.15$

31. Show that $\text{Arccos}(\cos x) = x$ if $0 \leq x \leq \pi$.

32. Show that $\text{Arcsin}(\sin x) = x$ if $-\frac{\pi}{2} \leq x \leq \frac{\pi}{2}$.

UNIT REVIEW

[2.1] *Verify that each of the following equations is an identity.*

1. $\cot x = \csc x \cos x$

2. $\dfrac{1}{1 + \tan^2 x} = \cos^2 x$

3. $\tan x + \cot x = \csc x \sec x$

4. $\dfrac{1 - \sin x}{\cos x} = \dfrac{\cos x}{1 + \sin x}$

5. $\sec x + \tan x = \dfrac{1}{\sec x - \tan x}$

6. $2 \cos^2 x - 1 = \cos^4 x - \sin^4 x$

[2.2] **7.** $\tan 2x = \dfrac{2 \sin x \cos x}{\cos^2 x - \sin^2 x}$

8. $\dfrac{\tan 2x}{2 \tan x} = \dfrac{\cot^2 x}{\cot^2 x - 1}$

9. Given that $\cos x = -\frac{2}{3}$, find:

(*a*) $\cos 2x, \quad \frac{\pi}{2} \leq x \leq \pi$

(*b*) $\sin 2x, \quad \pi \leq x \leq \frac{3\pi}{2}$

(*c*) $\sin \frac{1}{2}x, \quad 3\pi \leq x \leq \frac{7\pi}{2}$

(*d*) $\cos \frac{1}{2}x, \quad \frac{\pi}{2} \leq x \leq \pi$

10. Given that $\sin x = 0.6$, find an approximation for:

(*a*) $\sin 2x, \quad 0 \leq x \leq \frac{\pi}{2}$

(*b*) $\cos 2x, \quad \frac{3\pi}{2} \leq x \leq 2\pi$

(*c*) $\sin \frac{1}{2}x, \quad 0 \leq x \leq \frac{\pi}{2}$

(*d*) $\cos \frac{1}{2}x, \quad \frac{3\pi}{2} \leq x \leq 2\pi$

Find each function value by using appropriate reduction formulas.

11. $\sin \frac{5\pi}{4}$ **12.** $\tan \frac{7\pi}{6}$ **13.** $\cos \frac{11\pi}{6}$

14. $\tan 330°$ **15.** $\sin (-60°)$ **16.** $\cos (-45°)$

In Exercises 17 *to* 36, *when using the tables of function values, give results to the nearest reading in the table.*

[2.3] *Find the solution set of each equation in* (*a*) *radians and* (*b*) *degrees.*

17. $\sin x = \frac{1}{2}$ **18.** $\cot x = -\sqrt{3}$

19. $4 \cos x - 1 = 0$ **20.** $3 \sec x + 5 = 0$

Find the solution set of each equation in radians.

21. $\sqrt{3} \sin x + \cos x = 0$ **22.** $\cos^2 x - \sin^2 x = -1$

Find the solution set of each equation over the interval $0° \leq \theta < 360°$.

23. $(\tan \theta + 1)(2 \sin \theta - 1) = 0$ **24.** $\tan \theta \cos \theta - \cos \theta = 0$

Find the solution set of each equation over the interval $0^R \leq \alpha < 2\pi^R$.

25. $2 \sin^2 \alpha - \sin \alpha - 1 = 0$ **26.** $2 \tan^2 \alpha = \tan \alpha + 3$

[2.4] *Find the value of each expression in terms of rational multiples of* π.

27. $\text{Arccos} \frac{\sqrt{3}}{2}$ **28.** $\text{Sin}^{-1}\left(-\frac{\sqrt{3}}{2}\right)$

Find the value of each expression.

29. Arccos 0.7648 **30.** $\text{Tan}^{-1}(-0.2027)$

31. $\text{Cos}^{-1}\left(\sin \frac{\pi}{4}\right)$ **32.** $\text{Tan}^{-1}\left(\tan \frac{\pi}{6}\right)$

33. $\sin\left(\text{Arccos}\left(-\frac{1}{2}\right)\right)$

34. $\cos\,(\text{Sin}^{-1}\,0)$

Solve each equation over the intervals indicated in the definitions of each inverse function specifying the solution using inverse notation.

35. $3\tan\theta = 2$

36. $4\cos 3\theta = 3$

UNIT 3

TRIGONOMETRIC FORM OF COMPLEX NUMBERS; POLAR COORDINATES

There are several representations for complex numbers. Sometimes one form or another may be more useful. You probably have been introduced to the $a + bi$ form of a complex number. This form is considered in the Appendix beginning on page 54, and should be studied before starting Section 3.1 if you have not studied complex numbers in your earlier work in mathematics.

In this unit we shall consider the trigonometric form of a complex number. This is particularly useful in finding powers and roots of these numbers. In addition, we shall consider a new coordinate system which is closely related to some notions of complex numbers.

3.1 TRIGONOMETRIC FORM OF COMPLEX NUMBERS

Definitions and Notation*

Absolute value, or modulus, of a complex number $z = a + bi$

$|z|$ or $|a + bi|$

$|z| = |a + bi| = \sqrt{a^2 + b^2}$

An argument of a complex number $a + bi$

θ

An angle with initial side the positive x axis and terminal side the ray from the origin containing the graph of $a + bi$

* In this unit z will always represent a complex number.

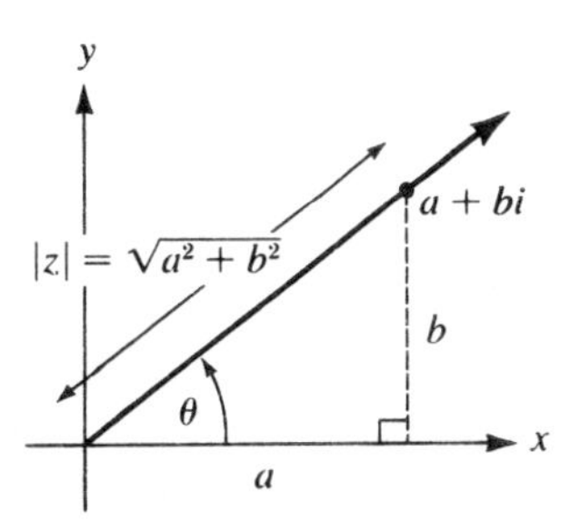

Example

For the complex number $z = \sqrt{3} + i$,
$a = \sqrt{3}$ and $b = 1$. Hence,

$$|z| = \sqrt{(\sqrt{3})^2 + 1^2} = 2$$

The argument is the angle θ in the figure.

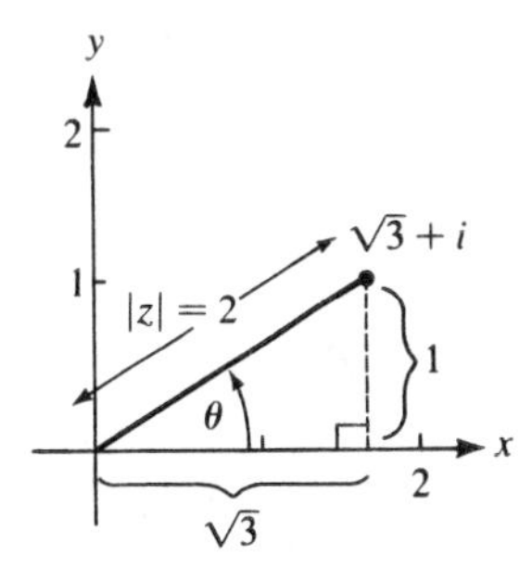

Trigonometric form of a complex number $z = r(\cos\theta + i \sin\theta)$, where $r = |z|$ and θ is an argument of z

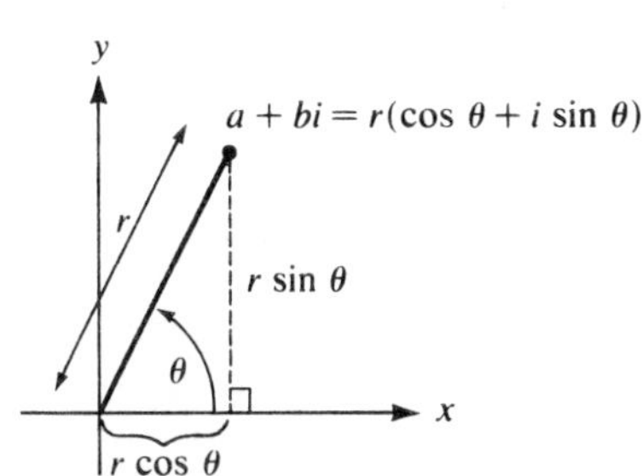

Example

For the complex number z whose argument is $\theta = 30°$ and with $|z| = r = 2$, $z = 2(\cos 30° + i \sin 30°)$.

cis θ An abbreviation for $\cos\theta + i \sin\theta$

Examples

a. $\cos 30° + i \sin 30° = \text{cis } 30°$
b. $r[\cos(-120°) + i \sin(-120°)] = r \text{ cis }(-120°)$
c. $r\left(\cos\frac{\pi^R}{3} + i \sin\frac{\pi^R}{3}\right) = r \text{ cis } \frac{\pi^R}{3}$

Properties

1. *If* θ *is an argument of* $a + bi$, *then so is* $\theta + k \cdot 360°$, $k \in J$.
2. *If* θ *is an argument of* $a + bi$, *then* $\tan \theta = b/a$, $a \neq 0$.
3. *If* $r(\cos \theta + i \sin \theta)$ *is a trigonometric form of a complex number, then so is* $r[\cos(\theta + k \cdot 360°) + i \sin(\theta + k \cdot 360°)]$, $k \in J$.
4. *If* $a + bi = r(\cos \theta + i \sin \theta)$, *then* $a = r \cos \theta$ *and* $b = r \sin \theta$.

If z_1 *and* z_2 *are complex numbers with*

$$z_1 = r_1(\cos \theta_1 + i \sin \theta_1) \quad \text{and} \quad z_2 = r_2(\cos \theta_2 + i \sin \theta_2),$$

5. $z_1 \cdot z_2 = r_1 r_2[\cos(\theta_1 + \theta_2) + i \sin(\theta_1 + \theta_2)] = r_1 r_2 \operatorname{cis}(\theta_1 + \theta_2)$
6. $\dfrac{z_1}{z_2} = \dfrac{r_1}{r_2}[\cos(\theta_1 - \theta_2) + i \sin(\theta_1 - \theta_2)] = \dfrac{r_1}{r_2} \operatorname{cis}(\theta_1 - \theta_2)$, $z_2 \neq 0 + 0i$

Exercises

Write without absolute-value notation.

Examples **a.** $|2 + 5i|$ **b.** $|-3|$

Solutions **a.** $|2 + 5i| = \sqrt{2^2 + 5^2} = \sqrt{29}$

b. $|-3| = |-3 + 0i| = \sqrt{(-3)^2 + 0^2} = 3$

1. $|3 + 2i|$ **2.** $|2 - 2i|$ **3.** $|4 - i|$

4. $|1 + i|$ **5.** $|4|$ **6.** $|-2|$

7. $|3i|$ **8.** $|-2i|$

Write each complex number in trigonometric form.

Example $1 + \sqrt{3}i$

Solution $r = |1 + \sqrt{3}i| = \sqrt{1^2 + (\sqrt{3})^2} = 2$; since $\tan \theta = \sqrt{3}/1$ and the graph of $1 + \sqrt{3}i$ is in Quadrant I, $\theta = 60°$. Hence,

$$1 + \sqrt{3}i = 2(\cos 60° + i \sin 60°)$$
$$= 2 \operatorname{cis} 60°$$

9. $3 + 3i$ **10.** $2 - 2i$ **11.** 5

12. $-7i$ **13.** $2\sqrt{3} - 2i$ **14.** $-3\sqrt{3} - 3i$

Write each complex number in the form $a + bi$.

Example $4 \text{ cis } 225°$

Solution Since $\cos 225° = -\sqrt{2}/2$ and $\sin 225° = -\sqrt{2}/2$, we have from Property 4,

$$a = r \cos \theta = 4\left(-\frac{\sqrt{2}}{2}\right) = -2\sqrt{2}$$

$$b = r \sin \theta = 4\left(-\frac{\sqrt{2}}{2}\right) = -2\sqrt{2}$$

and

$$a + bi = -2\sqrt{2} - 2\sqrt{2}i$$

15. $4 \text{ cis } 240°$ **16.** $3 \text{ cis } 300°$ **17.** $6 \text{ cis } (-30°)$
18. $5 \text{ cis } 180°$ **19.** $12 \text{ cis } 420°$ **20.** $10 \text{ cis } (-480°)$

For each given pair of complex numbers z_1 and z_2 in Exercises 21 to 26, find (a) $z_1 \cdot z_2$ and (b) z_1/z_2. Express each result in the form $a + bi$. Use Table II as needed.

Example $z_1 = 3 \text{ cis } 150°$ and $z_2 = 5 \text{ cis } 30°$

Solution (*a*)

$$\begin{aligned} 3 \text{ cis } 150° \cdot 5 \text{ cis } 30° &= 3 \cdot 5 \text{ cis } (150° + 30°) \\ &= 15 \text{ cis } 180° \\ &= 15(\cos 180° + i \sin 180°) \end{aligned}$$

Since $\cos 180° = -1$ and $\sin 180° = 0$,

$$15(\cos 180° + i \sin 180°) = 15(-1 + 0i) = -15$$

(*b*)

$$\begin{aligned} \frac{3 \text{ cis } 150°}{5 \text{ cis } 30°} &= \frac{3}{5}[\cos(150° - 30°) + i \sin (150° - 30°)] \\ &= \frac{3}{5}[\cos 120° + i \sin 120°] \end{aligned}$$

Since $\cos 120° = -1/2$ and $\sin 120° = \sqrt{3}/2$, we have

$$\frac{3}{5}(\cos 120° + i \sin 120°) = \frac{3}{5}\left(-\frac{1}{2} + \frac{\sqrt{3}}{2}i\right) = \frac{-3}{10} + \frac{3\sqrt{3}}{10}i$$

21. $z_1 = 3 \text{ cis } 90°$ and $z_2 = \sqrt{2} \text{ cis } 45°$
22. $z_1 = 4 \text{ cis } 30°$ and $z_2 = 2 \text{ cis } 60°$
23. $z_1 = 6 \text{ cis } 150°$ and $z_2 = 18 \text{ cis } 570°$

24. $z_1 = 14 \text{ cis } 210°$ and $z_2 = 2 \text{ cis } 120°$

25. $z_1 = -3 + i$ and $z_2 = -2 - 4i$

26. $z_1 = 2 + 3i$ and $z_2 = 2 + 3i$

27. Write $\left(-\frac{1}{2} - \frac{i\sqrt{3}}{2}\right)^3$ in the form $a + bi$.

28. Write $\left(-\frac{1}{2} + \frac{i\sqrt{3}}{2}\right)^3$ in the form $a + bi$.

29. Prove that the sum of two conjugate complex numbers is a real number.

30. Prove that the product of two conjugate complex numbers is a real number.

3.2 DE MOIVRE'S THEOREM: POWERS AND ROOTS

Definitions and Notation

nth root of z — w is an nth root of z if $w^n = z$

$\mathbf{z^0}$ — $z^0 = 1 + 0i$

$\mathbf{z^{-n}}$ — $z^{-n} = \frac{1}{z^n}$, for $n \in J$

Properties

1. *If z is a complex number,* $z = r \text{ cis } \theta$, *and* n *is an integer,* then

$$z^n = r^n \text{ cis } n\theta \qquad \textit{De Moivre's Theorem}$$

2. *If z is a complex number,* $z = r \text{ cis } \theta$, *and* $n \in N$, *then*

$$w_k = r^{1/n} \text{ cis } \left(\frac{\theta + k \cdot 360°}{n}\right), \qquad k \in J$$

gives the n *nth roots of z.*

Exercises

Write each of the given expressions as a complex number of the form $a + bi$. *Use Table II as necessary.*

Example $(\sqrt{3} + i)^7$

Solution For the modulus, we have $r = \sqrt{(\sqrt{3})^2 + 1^2} = 2$. We now find θ. From $\tan\theta = 1/\sqrt{3}$ and because the graph of $\sqrt{3} + i$ is in Quadrant I, we obtain $\theta = 30°$. Thus $(\sqrt{3} + i)^7 = (2 \text{ cis } 30°)^7$. Then, by De Moivre's Theorem,

$$(2 \text{ cis } 30°)^7 = 2^7 \text{ cis } (7 \cdot 30)° = 128 \text{ cis } 210°$$

Since $\cos 210° = -\sqrt{3}/2$ and $\sin 210° = -1/2$,

$$128(\cos 210° + i \sin 210°) = 128\left(-\frac{\sqrt{3}}{2} - \frac{1}{2}i\right) = -64\sqrt{3} - 64i$$

so

$$(\sqrt{3} + i)^7 = -64\sqrt{3} - 64i$$

1. $[2 \text{ cis } (-30°)]^7$ **2.** $(4 \text{ cis } 36°)^5$ **3.** $\left(-\frac{1}{2} + \frac{1}{2}\sqrt{3}\,i\right)^3$

4. $(1 + i)^{12}$ **5.** $(\sqrt{3} \text{ cis } 5°)^{12}$ **6.** $(\sqrt{2} \text{ cis } 12°)^{10}$

Example $(1 + i)^{-6}$

Solution Since $r = \sqrt{1^2 + 1^2} = \sqrt{2}$ and $\tan\theta = 1/1$ with the graph of $1 + i$ in Quadrant I, we have $\theta = 45°$. Hence,

$$(1 + i)^{-6} = (\sqrt{2} \text{ cis } 45°)^{-6}$$

By De Moivre's Theorem,

$$\begin{aligned}(\sqrt{2} \text{ cis } 45°)^{-6} &= (\sqrt{2})^{-6} \text{ cis } (-6 \cdot 45)° \\ &= \frac{1}{(\sqrt{2})^6} \text{ cis } (-270°) \\ &= \frac{1}{8}[\cos(-270°) + i \sin(-270°)]\end{aligned}$$

Since $\cos(-270°) = 0$ and $\sin(-270°) = 1$,

$$(1 + i)^{-6} = \frac{1}{8}(0 + i) = \frac{1}{8}i$$

7. $(\sqrt{3}-i)^{-5}$ **8.** $(1-i)^{-6}$ **9.** $(\text{cis } 60°)^{-3}$ **10.** $(\sqrt{2}\text{ cis } 30°)^{-7}$

Find all n of the nth roots of z. Leave the results in trigonometric form.

Example $z = 2 + 2\sqrt{3}i, \quad n = 4$

Solution Since $r = \sqrt{2^2 + (2\sqrt{3})^2} = 4$ and since the graph of $2 + 2\sqrt{3}i$ is in Quadrant I, $\theta = \text{Tan}^{-1}\, 2\sqrt{3}/2 = 60°$, we have

$$z = 2 + 2\sqrt{3}i = 4 \text{ cis } 60°$$

From Property 2, each number

$$w_k = 4^{1/4} \text{ cis } \left(\frac{60° + k \cdot 360°}{4}\right), \quad k \in J$$

is a fourth root of z. Taking $4^{1/4} = (4^{1/2})^{1/2} = \sqrt{2}$, and $k = 0, 1, 2,$ and 3 in turn, gives the roots

$$w_0 = \sqrt{2} \text{ cis } 15° \qquad w_1 = \sqrt{2} \text{ cis } 105°$$
$$w_2 = \sqrt{2} \text{ cis } 195° \qquad w_3 = \sqrt{2} \text{ cis } 285°$$

The substitution of any other integer for k will produce one of these four complex numbers.

11. $z = 32 \text{ cis } 45°, \quad n = 5$ **12.** $z = 27, \quad n = 3$

13. $z = -16\sqrt{3} + 16i, \quad n = 5$ **14.** $z = 1 - i, \quad n = 4$

15. $z = -i, \quad n = 6$ **16.** $z = 2 + 2\sqrt{3}i, \quad n = 3$

Solve each of the following equations over the set of complex numbers.

Example $x^4 = 2 + 2\sqrt{3}i$

Solution The numbers which are solutions of the given equation are the numbers x whose fourth powers are $2 + 2\sqrt{3}i$. Hence, we seek the fourth roots of $2 + 2\sqrt{3}i$ and we proceed as we did in the foregoing example to find the solutions

$$\sqrt{2} \text{ cis } 15°, \quad \sqrt{2} \text{ cis } 105°, \quad \sqrt{2} \text{ cis } 195°, \quad \sqrt{2} \text{ cis } 285°$$

17. $x^5 = 16 - 16\sqrt{3}i$ **18.** $x^3 + 4i = 4\sqrt{3}$

19. $x^7 + 1 = 0$ **20.** $x^7 - 1 = 0$

3.3 POLAR COORDINATES

Definition and Notation

Polar coordinates of a point A

(r, θ)

The components of an ordered pair (r,θ), where r is the distance from P (called the *pole*) to A, and θ is the angle from the x axis (taken through P) to $\overrightarrow{PA}$

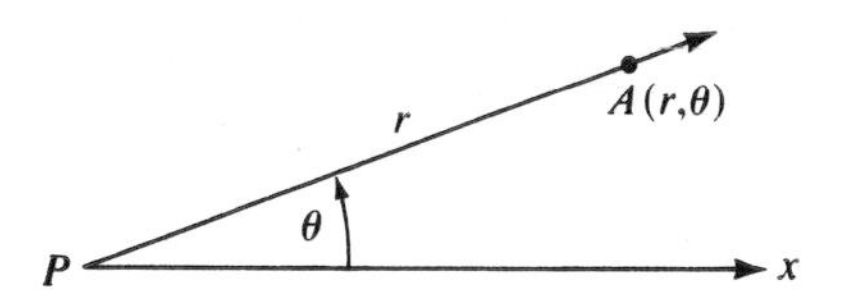

Properties

1. *Each point in the plane has infinitely many pairs of polar coordinates. If* (r,θ) *are polar coordinates of A, then so are*

$$(r, \theta + k \cdot 360°), \quad k \in J \quad \textit{and} \quad (-r, \theta + 180° + k \cdot 360°), \quad k \in J$$

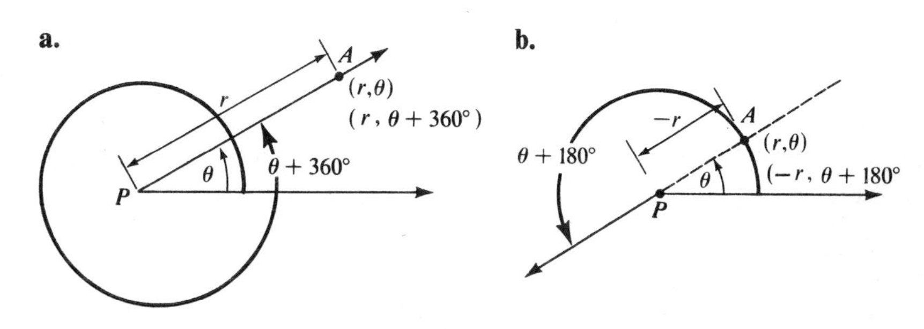

2. *Relationship between rectangular (Cartesian) and polar coordinates:*

   a. *To change polar to rectangular coordinates:*

$$x = r \cos \theta, \quad y = r \sin \theta$$

   b. *To change rectangular to polar coordinates:*

$$r = \pm\sqrt{x^2 + y^2},^{*} \quad \cos \theta = \frac{x}{\pm\sqrt{x^2 + y^2}}$$

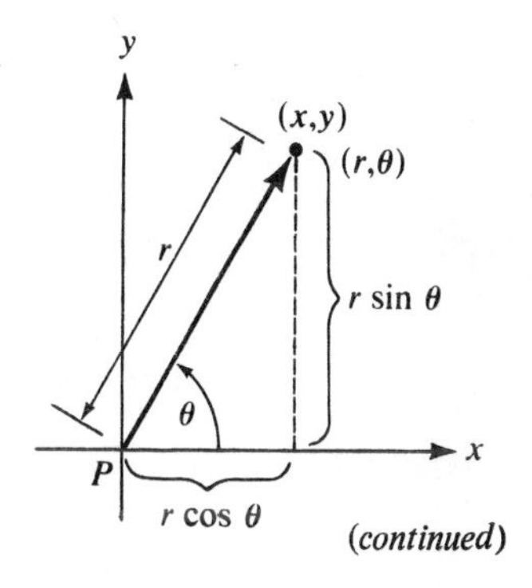

(continued)

* Note that the r defined in the context of polar coordinates can be positive or negative while the r defined in the context of complex numbers (page 41) is positive only.

$$\sin \theta = \frac{y}{\pm\sqrt{x^2+y^2}} \quad [(x,y) \neq (0,0)]$$

Also, $\tan \theta = y/x$ $(x \neq 0)$. *The particular problem determines whether the positive or negative sign is used with* $\sqrt{x^2+y^2}$.

Exercises

Find four sets of polar coordinates $(-360° < \theta \leq 360°)$ *for the point with polar coordinates as given.*

Example (3,390°)

Solution With positive values for r, two pairs of polar coordinates are (3,30°) and (3,−330°). Using negative values for r, we have (−3,210°) and (−3,−150°). The figures show these cases. Of course there are infinitely many other possible polar coordinates for the same point.

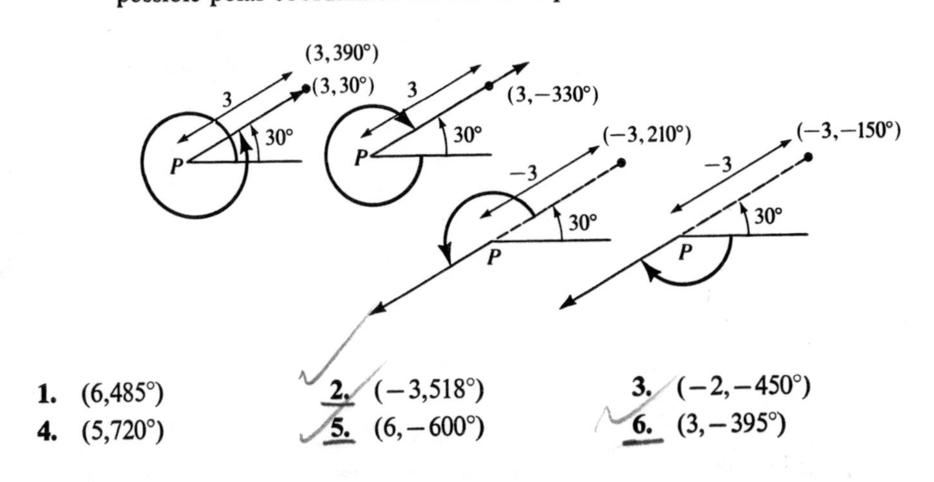

1. (6,485°) **2.** (−3,518°) **3.** (−2,−450°)
4. (5,720°) **5.** (6,−600°) **6.** (3,−395°)

Find the Cartesian coordinates of a point with polar coordinates as given.

Example (4,30°)

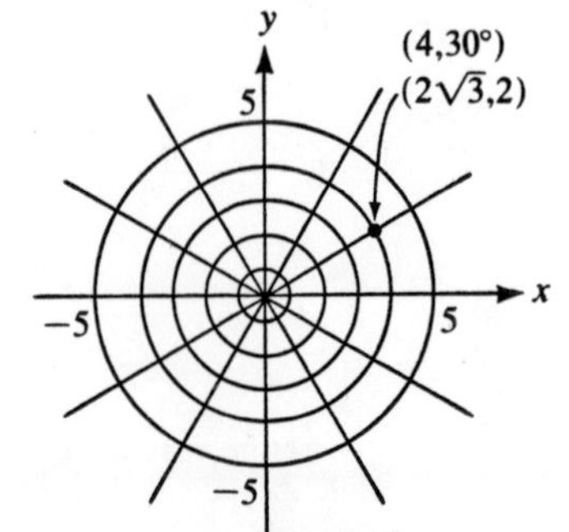

Solution Using Property 2a, we obtain

$$x = 4\cos 30° = 4 \cdot \frac{\sqrt{3}}{2} = 2\sqrt{3}$$

$$y = 4\sin 30° = 4 \cdot \frac{1}{2} = 2$$

Hence, the rectangular coordinates are $(2\sqrt{3},2)$.

7. $(5,45°)$ **8.** $(-3,30°)$ **9.** $\left(\frac{1}{2},330°\right)$

10. $\left(\frac{3}{4},225°\right)$ **11.** $(10,-135°)$ **12.** $(-6,-240°)$

Find two sets of polar coordinates, one involving an angle of positive measure and one an angle of negative measure, for the point with Cartesian coordinates as given.

Example $(7,-2)$

Solution By Property 2b, $r = \pm\sqrt{x^2+y^2} = \pm\sqrt{49+4} = \pm\sqrt{53}$. Choosing the positive sign, and noting that the point $(7,-2)$ is in the fourth quadrant, we have

$$\tan\theta = \frac{-2}{7}$$

from which

$$\theta \approx -16°00'$$

A pair of polar coordinates is

$$(\sqrt{53},-16°00')$$

where the given angle measure is an approximation. A second pair involving an angle of positive measure is

$$(\sqrt{53},344°00')$$

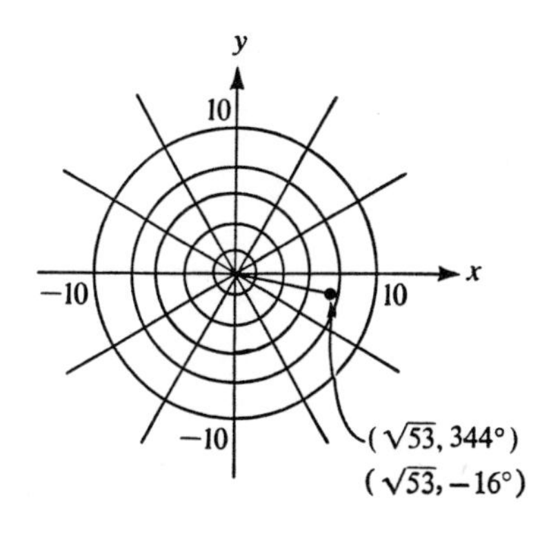

13. $(3\sqrt{2},3\sqrt{2})$ **14.** $\left(-\frac{\sqrt{3}}{2},\frac{1}{2}\right)$ **15.** $(-1,-\sqrt{3})$

16. $(0,-4)$ **17.** $(0,0)$ **18.** $(-6,0)$

Transform the given equation to an equation in polar form.

Example $x^2+y^2-2x+3=0$

Solution From Property 2a, we find that $x^2 = r^2\cos^2\theta$ and $y^2 = r^2\sin^2\theta$. Thus

$$r^2\cos^2\theta + r^2\sin^2\theta - 2r\cos\theta + 3 = 0$$

(continued)

from which

$$r^2(\cos^2\theta + \sin^2\theta) - 2r\cos\theta + 3 = 0$$
$$r^2 - 2r\cos\theta + 3 = 0$$

19. $x^2 + y^2 = 25$ **20.** $x = 3$ **21.** $y = -4$
22. $x^2 + y^2 - 4y = 0$ **23.** $x^2 + 9y^2 = 9$ **24.** $x^2 - 4y^2 = 4$

Transform the given equation to an equation in Cartesian form.

Example $r(1 - 2\cos\theta) = 3$

Solution Using Property 2b, we have

$$\pm\sqrt{x^2+y^2}\left[1 - 2\left(\frac{x}{\pm\sqrt{x^2+y^2}}\right)\right] = 3$$

from which

$$\pm\sqrt{x^2+y^2} - 2x = 3,$$
$$\pm\sqrt{x^2+y^2} = 2x + 3$$

Upon squaring each member, we have the equation

$$x^2 + y^2 = 4x^2 + 12x + 9$$

from which

$$3x^2 - y^2 + 12x + 9 = 0$$

25. $r = 5$ **26.** $r = 4\sin\theta$ **27.** $r = 9\cos\theta$
28. $r\cos\theta = 3$ **29.** $r(1 - \cos\theta) = 2$ **30.** $r(1 + \sin\theta) = 2$

31. Show by transformation of coordinates that the graph of $r = \sec^2\frac{\theta}{2}$ is a parabola. (*Hint:* An equation of the form $x = ay^2 + by + c$ is an equation of a parabola.)

32. Show by transformation of coordinates that the graph of $r = \csc^2\frac{\theta}{2}$ is a parabola.

Graph each equation.

Example $r = 4\sin\theta$

Solution First obtain some ordered pairs using 0°, 30°, 45°, 60°, 90°, 120°, 135°, etc. The tabulation of the data in the arrangement shown is helpful.

θ	$\sin\theta$	r or $4\sin\theta$	(r,θ)
0°	0	0	(0,0°)
30°	0.50	2.0	(2.0,30°)
45°	0.71	2.8	(2.8,45°)
60°	0.87	3.5	(3.5,60°)
90°	1.00	4.0	(4.0,90°)
120°	0.87	3.5	(3.5,120°)
135°	0.71	2.8	(2.8,135°)
150°	0.50	2.0	(2.0,150°)
180°	0	0	(0,180°)

Graph each ordered pair (r,θ) as in figure (*a*) and connect the points with a smooth curve as shown in figure (*b*). The values for r or $4\sin\theta$ for values of θ greater than 180° or less than 0° yield ordered pairs whose graphs will be the same as the graphs of the pairs already obtained.

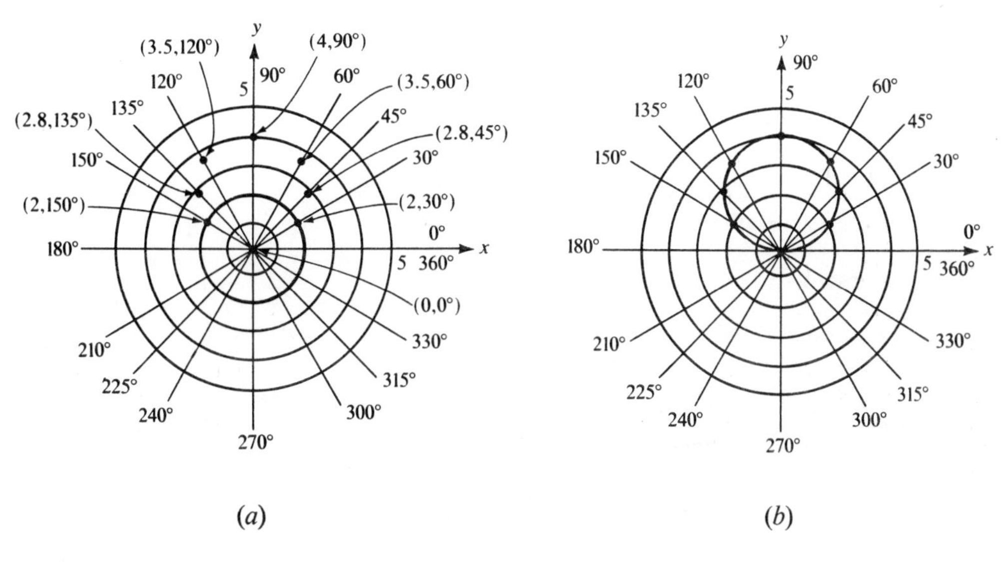

(*a*) (*b*)

33. $r = 9\sin\theta$

34. $r = 9\cos\theta$

35. $r = 1 + 2\cos\theta$

36. $r = 1 + 2\sin\theta$

37. $r = 1 - \sin\theta$

38. $r = 1 - \cos\theta$

39. $r = 4\sin 3\theta$

40. $r = \dfrac{2}{1 - \sin\theta}$

UNIT REVIEW

[3.1] *Write without absolute-value notation.*

1. $|3-4i|$ **2.** $|-5|$ **3.** $|2+3i|$ **4.** $|-3i|$

Write each complex number in trigonometric form.

5. $3-3i$ **6.** $-2i$

Write each complex number in the form $a+bi$.

7. $2 \text{ cis } 300°$ **8.** $6 \text{ cis } (-60°)$

For each given pair of complex numbers, z_1 and z_2, find (a) $z_1 \cdot z_2$ and (b) z_1/z_2. Express each result in the form $a+bi$.

9. $z_1 = 4 \text{ cis } 60°,\quad z_2 = 2 \text{ cis } 60°$ **10.** $z_1 = 3-i,\quad z_2 = 2+4i$

[3.2] *Write each given expression as a complex number in the form $a+bi$.*

11. $[2 \text{ cis } (-60°)]^5$ **12.** $(1-i)^{10}$

13. $(\sqrt{3}+i)^4$ **14.** $(-1+i)^{-6}$

Find all n of the nth roots of z. Leave the results in trigonometric form.

15. $z = 16 \text{ cis } 60°,\quad n=4$ **16.** $z = 16 + 16\sqrt{3}\,i,\quad n=5$

Solve each equation over the set of complex numbers.

17. $x^5 = 16\sqrt{3} + 16i$ **18.** $x^6 - 1 = 0$

[3.3] *Find four sets of polar coordinates $(-360° < \theta \leq 360°)$ for the point with polar coordinates as given.*

19. $(4, 500°)$ **20.** $(-3, -700°)$

Find the Cartesian coordinates of a point with polar coordinates as given.

21. $(-2, 60°)$

22. $(6, -225°)$

Find two sets of polar coordinates, one involving an angle of positive measure and one an angle of negative measure, for the point with Cartesian coordinates as given.

23. $(2\sqrt{2}, 2\sqrt{2})$

24. $(1, -\sqrt{3})$

Transform the given equation to an equation in polar form.

25. $9x^2 + y^2 = 9$

26. $9x^2 - y^2 = 9$

Transform the given equation to an equation in Cartesian form.

27. $r = 2$

28. $r + r \cos \theta = 2$

Graph each equation.

29. $r = 2 \sin \theta$

30. $r = 1 + \sin \theta$

APPENDIX

PRELIMINARY CONCEPTS OF COMPLEX NUMBERS*

Some equations do not have solutions in the set R of real numbers. For example,

$$x^2 = -9$$

and

$$x^2 + 2x + 5 = 0 \quad \text{equivalent to} \quad (x + 1)^2 = -4$$

have no solutions in R. In general if $b > 0$, then $x^2 = -b$ has no real-number solution because there is no real number whose square is negative. Hence any symbol of the form $\sqrt{-b}$, $b \in R$, $b > 0$ can not represent a real number. The set C of complex numbers includes members whose squares are negative real numbers, and also contains members that can be identified with the members of the set of real numbers and provides solutions for all polynomial equations in one variable.

A.1 SUMS AND DIFFERENCES

Definitions and Notation

Pure imaginary number A number whose square is a negative real number; $\sqrt{-b}\sqrt{-b} = -b$

$\sqrt{-b}$,

$b \in R$, $b > 0$

* From Volume 6 of this series by Charles Carico.

Examples

a. $\sqrt{-4}$, $\sqrt{-3}$, and $\sqrt{-1}$ are imaginary numbers
b. $\sqrt{-4}\sqrt{-4} = -4$, $\sqrt{-3}\sqrt{-3} = -3$, and $\sqrt{-1}\sqrt{-1} = -1$
c. $\sqrt{x}$ is a pure imaginary number if $x < 0$

i — $i = \sqrt{-1}$; $i \cdot i = i^2 = -1$

$i\sqrt{b}$ or $\sqrt{b}\,i$, $b > 0$ — $i\sqrt{b} = \sqrt{-1}\sqrt{b} = \sqrt{-b}$, $b > 0$

Examples

a. $\sqrt{-4} = \sqrt{-1}\sqrt{4} = i\sqrt{4} = 2i$
b. $\sqrt{-3} = \sqrt{-1}\sqrt{3} = i\sqrt{3} = \sqrt{3}\,i$

Set of complex numbers — The set whose members include all the members of the set of real numbers, and all possible sums and products of real numbers and pure imaginary numbers

Imaginary numbers — The members of the set of complex numbers that are not real numbers

Standard forms for a complex number — $a + bi$ or $a + ib$, $a, b \in R$

z — A variable that is often used to represent an unspecified element in the set of complex numbers; $z = a + bi$, $a, b \in R$*

$-z$ — The negative of z; if $z = a + bi$ then $-z = -a - bi$

Examples

a. $2 + 3i$, $-3 - i\sqrt{5}$, and $4 - \sqrt{2}\,i$ are in standard form
b. $0 + 4i$ or $4i$ is in standard form
c. $5 + 0i$ or 5 is in standard form
d. The negative of $2 + 3i$ is $-(2 + 3i)$ or $-2 - 3i$.
e. If $z = 4 - 2i$, then $-z = -(4 - 2i) = -4 + 2i$.

* In this volume z will always represent a complex number.

C The set of complex numbers

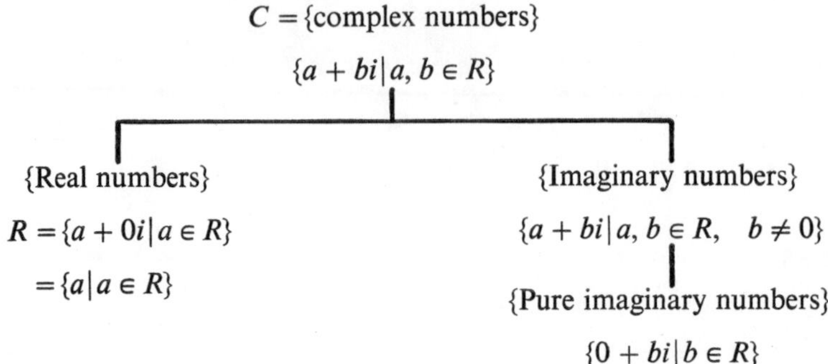

Properties

1. $(a_1 + b_1 i) + (a_2 + b_2 i) = (a_1 + a_2) + (b_1 + b_2)i$
2. $z_1 + z_2 \in C$ — *Closure law for addition in C*
3. $z_1 + z_2 = z_2 + z_1$ — *Commutative law of addition in C*
4. $z_1 + (z_2 + z_3) = (z_1 + z_2) + z_3$ — *Associative law of addition in C*
5. $z_1 - z_2 = z_1 + (-z_2)$
6. $(a_1 + b_1 i) - (a_2 + b_2 i) = (a_1 - a_2) + (b_1 - b_2)i$

Exercises

Write each expression in the form a + bi, a + ib, or bi.

Examples **a.** $\sqrt{-16}$ **b.** $2 - 3\sqrt{-16}$

Solutions **a.** $\sqrt{-16} = \sqrt{-1 \cdot 16} = \sqrt{-1}\sqrt{16} = 4i$

b. $2 - 3\sqrt{-16} = 2 - 3\sqrt{-1 \cdot 16} = 2 - 3\sqrt{-1}\sqrt{16} = 2 - 12i$

1. $\sqrt{-4}$ **2.** $\sqrt{-9}$ **3.** $\sqrt{-32}$

4. $\sqrt{-50}$ **5.** $3\sqrt{-8}$ **6.** $4\sqrt{-18}$

7. $4 + 2\sqrt{-1}$ **8.** $5 - 3\sqrt{-1}$ **9.** $3\sqrt{-50} + 2$

10. $5\sqrt{-12} - 1$ **11.** $\sqrt{4} + \sqrt{-4}$ **12.** $\sqrt{20} - \sqrt{-20}$

13. $\frac{1}{3}\sqrt{-25}$ **14.** $\frac{3}{4}\sqrt{-49}$ **15.** $\frac{\sqrt{-72}}{5}$

16. $\frac{\sqrt{-12}}{3}$ **17.** $\frac{-3\sqrt{-50}}{2}$ **18.** $\frac{-4\sqrt{-75}}{7}$

Examples

a. $(3 - i) - (6 + 2i)$

b. $(4 - 3\sqrt{-1}) + (2 + \sqrt{-1})$

Solutions

a. $(3 - i) - (6 + 2i)$
$= (3 - 6) + (-1 - 2)i$
$= -3 - 3i$

b. $(4 - 3\sqrt{-1}) + (2 + \sqrt{-1})$
$= (4 - 3i) + (2 + i)$
$= (4 + 2) + (-3 + 1)i$
$= 6 - 2i$

19. $(5 + i) + (3 - 3i)$

20. $(2 - 5i) + (4 + i)$

21. $(2 - 3i) + (4 + i)$

22. $(6 + i) + (3 - 2i)$

23. $i + (2 - 3i)$

24. $(3i - 1) + 5i$

25. $(4 - 3i) - (2 - 3i)$

26. $(5 + 3i) - (3 - 2i)$

27. $6 - (3 - 3i)$

28. $2i - (4 + 3i)$

29. $(3 - \sqrt{-1}) + (2 + 3\sqrt{-1})$

30. $(4 + 2\sqrt{-1}) + (5 - 3\sqrt{-1})$

31. $(3 - \sqrt{-4}) - (5 - \sqrt{-9})$

32. $(1 + \sqrt{-16}) - (2 - \sqrt{-25})$

33. $(4 + \sqrt{-8}) + (2 - \sqrt{-2})$

34. $(5 - \sqrt{-12}) + (3 + \sqrt{-27})$

35. $(2 - \sqrt{-3}) - (5 + \sqrt{-12})$

36. $(4 + \sqrt{-20}) - (1 - \sqrt{-5})$

A.2 PRODUCTS AND QUOTIENTS

Definition and Notation

Conjugate of $z = a + bi$

$\bar{z}$

The complex number $\bar{z} = a - bi$

Examples

a. The conjugate of $2 - 3i$ is $2 + 3i$, and the conjugate of $6 + i$ is $6 - i$.

b. If $z = -5 - 2i$, then $\bar{z} = -5 + 2i$ and if $z = -1 + 3i$, then $\bar{z} = -1 - 3i$.

Properties

1. $(a_1 + b_1 i)(a_2 + b_2 i) = a_1 a_2 + a_1 b_2 i + a_2 b_1 i + b_1 b_2 i^2$
$= (a_1 a_2 - b_1 b_2) + (a_1 b_2 + a_2 b_1)i$

2. $z_1 \cdot z_2 \in C$ *Closure law for multiplication in C*

3. $z_1 \cdot z_2 = z_2 \cdot z_1$ — *Commutative law of multiplication in C*
4. $z_1 \cdot (z_2 \cdot z_3) = (z_1 \cdot z_2) \cdot z_3$ — *Associative law of multiplication in C*
5. $z_1 \cdot (z_2 + z_3) = z_1 \cdot z_2 + z_1 \cdot z_3$ — *Distributive law in C*
6. $\frac{z}{c} = \frac{1}{c} \cdot z \quad (c \neq 0)$
7. $\frac{z_1}{z_2} = \frac{z_1 z_3}{z_2 z_3} \quad (z_2, z_3 \neq 0 + 0i)$ — *Fundamental principle of fractions*

Exercises

Write each expression in the form $a + bi$ or $a - bi$.

Examples **a.** $i(3 + 2i)$ **b.** $(2 - \sqrt{-1})(3 + \sqrt{-2})$

Solutions **a.** $i(3 + 2i) = 3i + 2i^2 = -2 + 3i$

b. $(2 - \sqrt{-1})(3 + \sqrt{-2}) = (2 - i)(3 + 2i) = 6 + 4i - 3i - 2i^2 = 6 + i + 2 = 8 + i$

1. $2(5 + i)$ **2.** $4(3 - 2i)$ **3.** $i(2 - i)$

4. $i(4 + i)$ **5.** $-3i(1 + 2i)$ **6.** $-2i(5 - 4i)$

7. $(5 - 2i)(5 + 2i)$ **8.** $(1 + 5i)(1 - 5i)$ **9.** $(3 + i\sqrt{2})(3 - i\sqrt{2})$

10. $(6 - i\sqrt{3})(6 + i\sqrt{3})$ **11.** $2(5 - i) - 3(2 + i)$ **12.** $4(2 + 3i) - 2(3 - 2i)$

13. $(1 + i)^2 - (1 - i)^2$ **14.** $(1 - i)^2 - (1 + i)^2$

15. $\sqrt{-3}(2 + \sqrt{-3})$ **16.** $\sqrt{-5}(4 - \sqrt{-5})$

17. $(3 + 2\sqrt{-5})(3 - \sqrt{-5})$ **18.** $(3 + \sqrt{-3})(4 + 2\sqrt{-3})$

19. $(2 - 4\sqrt{-3})(2 + 4\sqrt{-3})$ **20.** $(1 - 3\sqrt{-2})(1 + 3\sqrt{-2})$

Examples **a.** $\frac{-2}{i}$ **b.** $\frac{1}{2 - \sqrt{-9}}$

Solutions For a, multiply numerator and denominator by i. For b, multiply numerator and denominator by the conjugate of the denominator.

a. $\frac{-2}{i} = \frac{-2(i)}{i(i)} = \frac{-2i}{i^2} = \frac{-2i}{-1} = 2i$

b. $\frac{1}{2 - \sqrt{-9}} = \frac{1(2 + 3i)}{(2 - 3i)(2 + 3i)} = \frac{2 + 3i}{4 - 9i^2} = \frac{2 + 3i}{13} = \frac{2}{13} + \frac{3}{13}i$

21. $\frac{1}{i}$ 22. $\frac{-3}{i}$ 23. $\frac{-3}{\sqrt{-4}}$ 24. $\frac{7}{\sqrt{-9}}$

25. $\frac{-2}{1-i}$ 26. $\frac{3}{2+i}$ 27. $\frac{3}{3+2i}$ 28. $\frac{-5}{2-i}$

29. $\frac{i}{2+3i}$ 30. $\frac{3i}{1-i}$ 31. $\frac{\sqrt{-1}-1}{\sqrt{-1}+1}$ 32. $\frac{1+\sqrt{-4}}{1-\sqrt{-4}}$

Simplify.

Examples **a.** i^7 **b.** i^{21}

Solutions **a.** $i^7 = i^4 \cdot i^2 \cdot i = 1 \cdot (-1) \cdot i = -i$

b. $i^{21} = i^4 \cdot i^4 \cdot i^4 \cdot i^4 \cdot i^4 \cdot i = 1 \cdot 1 \cdot 1 \cdot 1 \cdot 1 \cdot 1 \cdot i = i$

33. (*a*) i^6 (*b*) i^{12} 34. (*a*) i^{15} (*b*) i^{25}

A.3 GRAPHICAL REPRESENTATION

Definitions and Notation

Complex plane A plane on which complex numbers are represented

Real axis Horizontal axis of the complex plane

Imaginary axis Vertical axis of the complex plane

(a,b) $(a,b) = a + bi$

Examples

a. $(-2,-3) = -2 - 3i$ and $(4,0) = 4 + 0i = 4$

b. $2 + 5i = (2,5)$ and $-6i = 0 - 6i = (0,-6)$

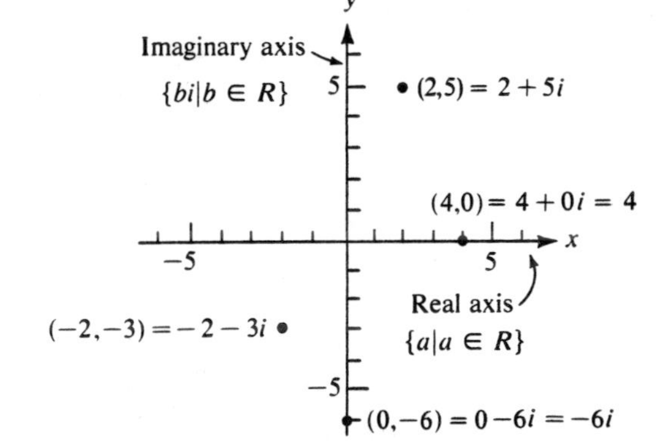

Exercises

Express each complex number (a,b) in the form $a + bi$.

Examples **a.** $(5,3)$ **b.** $(3,0)$

Solutions **a.** $(5,3) = 5 + 3i$ **b.** $(3,0) = 3 + 0i = 3$

1. $(2,6)$ **2.** $(-3,4)$ **3.** $(5,-2)$ **4.** $(0,6)$
5. $(-7,-3)$ **6.** $(-3,2)$ **7.** $(4,0)$ **8.** $(0,0)$

Express each complex number as an ordered pair.

Examples **a.** $4 - 5i$ **b.** $-2i$

Solutions **a.** $4 - 5i = (4,-5)$ **b.** $-2i = 0 - 2i = (0,-2)$

9. $2 + 3i$ **10.** $4 - 2i$ **11.** $-3 + i$ **12.** $-6 - 3i$
13. $4i$ **14.** 0 **15.** 7 **16.** $-i$

Graph each complex number, its conjugate, and its negative.

Examples **a.** $2 - 3i$ **b.** $(-3,2)$

Solutions **a.** The conjugate is $2 + 3i$; the negative is $-2 + 3i$.
b. The conjugate is $(-3,-2)$; the negative is $(3,-2)$.

a.

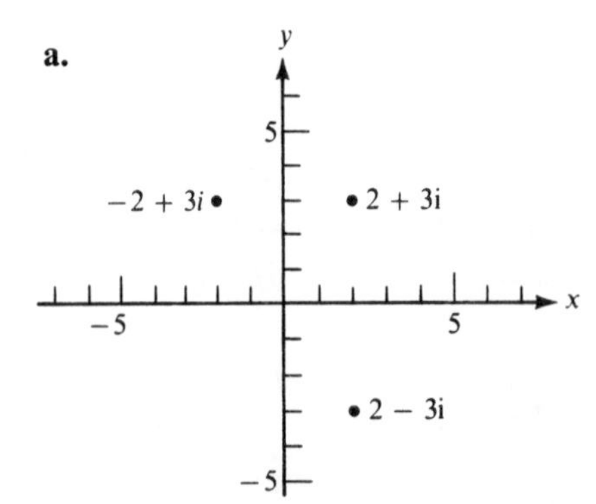

b.

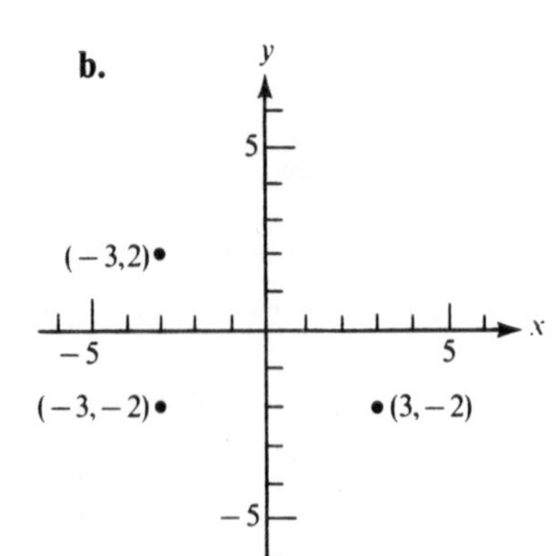

17. $2 + 3i$ **18.** $-3 + 4i$ **19.** $4 - i$ **20.** $-2 - i$
21. $4i$ **22.** $-3i$ **23.** $(4,-2)$ **24.** $(0,3)$

Graph (*a*) *the given complex number* $a + bi$,
(*b*) *the product* $i \cdot (a + bi)$,
(*c*) *the quotient* $\frac{a + bi}{i}$.
(*d*) *Draw arrows from the origin to the graph of each complex number.*

Example $3 + 2i$

Solution

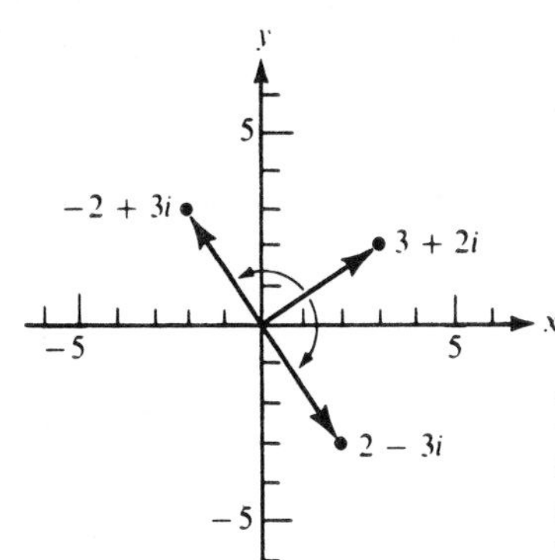

$$i \cdot (3 + 2i) = 3i + 2i^2$$
$$= 3i + 2(-1)$$
$$= -2 + 3i;$$

$$\frac{3 + 2i}{i} = \frac{(3 + 2i) \cdot i}{i \cdot i} = \frac{3i + 2i^2}{-1}$$
$$= \frac{3i + 2(-1)}{-1} = 2 - 3i$$

Note that multiplication of the complex number by i results in a 90° counterclockwise rotation of the arrow to $3 + 2i$. The division of the complex number by i results in a 90° clockwise rotation to $3 + 2i$.

25. $2 + 3i$ **26.** $-4 + 2i$ **27.** $-3 - 3i$
28. $4 - 3i$ **29.** $2i$ **30.** 4

APPENDIX REVIEW

[A.1] *Write each expression in the form* $a + bi$.

1. $\sqrt{-36}$ **2.** $\sqrt{-32}$ **3.** $2\sqrt{-8}$ **4.** $\frac{2}{3}\sqrt{-18}$

5. $(2 - i) + (5 + 6i)$ **6.** $(3 + 2i) - (4 - i)$

7. $(2 - \sqrt{-8}) + \sqrt{-2}$ **8.** $\sqrt{-5} - (4 - \sqrt{-20})$

[A.2] *Write each expression in the form* $a + bi$.

9. $3(2 - i)$ **10.** $i(5 + 2i)$ **11.** $(5 - 2i)(3 + i)$

12. $(2 + i)(2 - i)$ **13.** $\sqrt{-2}(1 + \sqrt{-2})$ **14.** $\sqrt{-3}(3 - \sqrt{-12})$

15. $\frac{-4}{i}$ **16.** $\frac{8}{\sqrt{-16}}$ **17.** $\frac{2}{1 - 2i}$ **18.** $\frac{\sqrt{-3}}{2 + \sqrt{-3}}$

Simplify

19. i^5

20. i^{19}

[A.3] *Express each complex number in the form* $a + bi$.

21. $(3, -5)$

22. $(-2, 2)$

Express each complex number as an ordered pair.

23. $5 - i$

24. $-3 + 4i$

Graph each complex number, its conjugate, and its negative.

25. $-4 + 2i$

26. $(0, -3)$

Table I Squares, Square Roots, and Prime Factors

No.	Sq.	Sq. Rt.	Factors	No.	Sq.	Sq. Rt.	Factors
1	1	1.000		51	2,601	7.141	$3 \cdot 17$
2	4	1.414	2	52	2,704	7.211	$2^2 \cdot 13$
3	9	1.732	3	53	2,809	7.280	53
4	16	2.000	2^2	54	2,916	7.348	$2 \cdot 3^3$
5	25	2.236	5	55	3,025	7.416	$5 \cdot 11$
6	36	2.449	$2 \cdot 3$	56	3,136	7.483	$2^3 \cdot 7$
7	49	2.646	7	57	3,249	7.550	$3 \cdot 19$
8	64	2.828	2^3	58	3,364	7.616	$2 \cdot 29$
9	81	3.000	3^2	59	3,481	7.681	59
10	100	3.162	$2 \cdot 5$	60	3,600	7.746	$2^2 \cdot 3 \cdot 5$
11	121	3.317	11	61	3,721	7.810	61
12	144	3.464	$2^2 \cdot 3$	62	3,844	7.874	$2 \cdot 31$
13	169	3.606	13	63	3,969	7.937	$3^2 \cdot 7$
14	196	3.742	$2 \cdot 7$	64	4,096	8.000	2^6
15	225	3.873	$3 \cdot 5$	65	4,225	8.062	$5 \cdot 13$
16	256	4.000	2^4	66	4,356	8.124	$2 \cdot 3 \cdot 11$
17	289	4.123	17	67	4,489	8.185	67
18	324	4.243	$2 \cdot 3^2$	68	4,624	8.246	$2^2 \cdot 17$
19	361	4.359	19	69	4,761	8.307	$3 \cdot 23$
20	400	4.472	$2^2 \cdot 5$	70	4,900	8.367	$2 \cdot 5 \cdot 7$
21	441	4.583	$3 \cdot 7$	71	5,041	8.426	71
22	484	4.690	$2 \cdot 11$	72	5,184	8.485	$2^3 \cdot 3^2$
23	529	4.796	23	73	5,329	8.544	73
24	576	4.899	$2^3 \cdot 3$	74	5,476	8.602	$2 \cdot 37$
25	625	5.000	5^2	75	5,625	8.660	$3 \cdot 5^2$
26	676	5.099	$2 \cdot 13$	76	5,776	8.718	$2^2 \cdot 19$
27	729	5.196	3^3	77	5,929	8.775	$7 \cdot 11$
28	784	5.292	$2^2 \cdot 7$	78	6,084	8.832	$2 \cdot 3 \cdot 13$
29	841	5.385	29	79	6,241	8.888	79
30	900	5.477	$2 \cdot 3 \cdot 5$	80	6,400	8.944	$2^4 \cdot 5$
31	961	5.568	31	81	6,561	9.000	3^4
32	1,024	5.657	2^5	82	6,724	9.055	$2 \cdot 41$
33	1,089	5.745	$3 \cdot 11$	83	6,889	9.110	83
34	1,156	5.831	$2 \cdot 17$	84	7,056	9.165	$2^2 \cdot 3 \cdot 7$
35	1,225	5.916	$5 \cdot 7$	85	7,225	9.220	$5 \cdot 17$
36	1,296	6.000	$2^2 \cdot 3^2$	86	7,396	9.274	$2 \cdot 43$
37	1,369	6.083	37	87	7,569	9.327	$3 \cdot 29$
38	1,444	6.164	$2 \cdot 19$	88	7,744	9.381	$2^3 \cdot 11$
39	1,521	6.245	$3 \cdot 13$	89	7,921	9.434	89
40	1,600	6.325	$2^3 \cdot 5$	90	8,100	9.487	$2 \cdot 3^2 \cdot 5$
41	1,681	6.403	41	91	8,281	9.539	$7 \cdot 13$
42	1,764	6.481	$2 \cdot 3 \cdot 7$	92	8,464	9.592	$2^2 \cdot 23$
43	1,849	6.557	43	93	8,649	9.644	$3 \cdot 31$
44	1,936	6.633	$2^2 \cdot 11$	94	8,836	9.695	$2 \cdot 47$
45	2,025	6.708	$3^2 \cdot 5$	95	9,025	9.747	$5 \cdot 19$
46	2,116	6.782	$2 \cdot 23$	96	9,216	9.798	$2^5 \cdot 3$
47	2,209	6.856	47	97	9,409	9.849	97
48	2,304	6.928	$2^4 \cdot 3$	98	9,604	9.899	$2 \cdot 7^2$
49	2,401	7.000	7^2	99	9,801	9.950	$3^2 \cdot 11$
50	2,500	7.071	$2 \cdot 5^2$	100	10,000	10.000	$2^2 \cdot 5^2$

Table II Values of Trigonometric Functions (θ in Degrees)

Angle θ	$\sin\theta$	$\csc\theta$	$\tan\theta$	$\cot\theta$	$\sec\theta$	$\cos\theta$	
0° 00′	.0000	No value	.0000	No value	1.000	1.0000	90° 00′
10	029	343.8	029	343.8	000	000	50
20	058	171.9	058	171.9	000	000	40
30	087	114.6	087	114.6	000	1.0000	30
40	116	85.95	116	85.94	000	.9999	20
50	145	68.76	145	68.75	000	999	10
1° 00′	.0175	57.30	.0175	57.29	1.000	.9998	89° 00′
10	204	49.11	204	49.10	000	998	50
20	233	42.98	233	42.96	000	997	40
30	262	38.20	262	38.19	000	997	30
40	291	34.38	291	34.37	000	996	20
50	320	31.26	320	31.24	001	995	10
2° 00′	.0349	28.65	.0349	28.64	1.001	.9994	88° 00′
10	378	26.45	378	26.43	001	993	50
20	407	24.56	407	24.54	001	992	40
30	436	22.93	437	22.90	001	990	30
40	465	21.49	466	21.47	001	989	20
50	494	20.23	495	20.21	001	988	10
3° 00′	.0523	19.11	.0524	19.08	1.001	.9986	87° 00′
10	552	18.10	553	18.07	002	985	50
20	581	17.20	582	17.17	002	983	40
30	610	16.38	612	16.35	002	981	30
40	640	15.64	641	15.60	002	980	20
50	669	14.96	670	14.92	002	978	10
4° 00′	.0698	14.34	.0699	14.30	1.002	.9976	86° 00′
10	727	13.76	729	13.73	003	974	50
20	756	13.23	758	13.20	003	971	40
30	785	12.75	787	12.71	003	969	30
40	814	12.29	816	12.25	003	967	20
50	843	11.87	846	11.83	004	964	10
5° 00′	.0872	11.47	.0875	11.43	1.004	.9962	85° 00′
10	901	11.10	904	11.06	004	959	50
20	929	10.76	934	10.71	004	957	40
30	958	10.43	963	10.39	005	954	30
40	.0987	10.13	.0992	10.08	005	951	20
50	.1016	9.839	.1022	9.788	005	948	10
6° 00′	.1045	9.567	.1051	9.514	1.006	.9945	84° 00′
10	074	9.309	080	9.255	006	942	50
20	103	9.065	110	9.010	006	939	40
30	132	8.834	139	8.777	006	936	30
40	161	8.614	169	8.556	007	932	20
50	190	8.405	198	8.345	007	929	10
7° 00′	.1219	8.206	.1228	8.144	1.008	.9925	83° 00′
10	248	8.016	257	7.953	008	922	50
20	276	7.834	287	7.770	008	918	40
30	305	7.661	317	7.596	009	914	30
40	334	7.496	346	7.429	009	911	20
50	363	7.337	376	7.269	009	907	10
8° 00′	.1392	7.185	.1405	7.115	1.010	.9903	82° 00
	$\cos\theta$	$\sec\theta$	$\cot\theta$	$\tan\theta$	$\csc\theta$	$\sin\theta$	Angle θ

Table II (continued)

Angle θ	sin θ	csc θ	tan θ	cot θ	sec θ	cos θ	
8° 00′	.1392	7.185	.1405	7.115	1.010	.9903	82° 00′
10	421	7.040	435	6.968	010	899	50
20	449	6.900	465	827	011	894	40
30	478	765	495	691	011	890	30
40	507	636	524	561	012	886	20
50	536	512	554	435	012	881	10
9° 00′	.1564	6.392	.1584	6.314	1.012	.9877	81° 00′
10	593	277	614	197	013	872	50
20	622	166	644	6.084	013	868	40
30	650	6.059	673	5.976	014	863	30
40	679	5.955	703	871	014	858	20
50	708	855	733	769	015	853	10
10° 00′	.1736	5.759	.1763	5.671	1.015	.9848	80° 00′
10	765	665	793	576	016	843	50
20	794	575	823	485	016	838	40
30	822	487	853	396	017	833	30
40	851	403	883	309	018	827	20
50	880	320	914	226	018	822	10
11° 00′	.1908	5.241	.1944	5.145	1.019	.9816	79° 00′
10	937	164	.1974	5.066	019	811	50
20	965	089	.2004	4.989	020	805	40
30	.1994	5.016	035	915	020	799	30
40	.2022	4.945	065	843	021	793	20
50	051	876	095	773	022	787	10
12° 00′	.2079	4.810	.2126	4.705	1.022	.9781	78° 00′
10	108	745	156	638	023	775	50
20	136	682	186	574	024	769	40
30	164	620	217	511	024	763	30
40	193	560	247	449	025	757	20
50	221	502	278	390	026	750	10
13° 00′	.2250	4.445	.2309	4.331	1.026	.9744	77° 00′
10	278	390	339	275	027	737	50
20	306	336	370	219	028	730	40
30	334	284	401	165	028	724	30
40	363	232	432	113	029	717	20
50	391	182	462	061	030	710	10
14° 00′	.2419	4.134	.2493	4.011	1.031	.9703	76° 00′
10	447	086	524	3.962	031	696	50
20	476	4.039	555	914	032	689	40
30	504	3.994	586	867	033	681	30
40	532	950	617	821	034	674	20
50	560	906	648	776	034	667	10
15° 00′	.2588	3.864	.2679	3.732	1.035	.9659	75° 00′
10	616	822	711	689	036	652	50
20	644	782	742	647	037	644	40
30	672	742	773	606	038	636	30
40	700	703	805	566	039	628	20
50	728	665	836	526	039	621	10
16° 00′	.2756	3.628	.2867	3.487	1.040	.9613	74° 00′
	cos θ	sec θ	cot θ	tan θ	csc θ	sin θ	Angle θ

Table II (continued)

Angle θ	sin θ	csc θ	tan θ	cot θ	sec θ	cos θ	
16° 00′	.2756	3.628	.2867	3.487	1.040	.9613	74° 00′
10	784	592	899	450	041	605	50
20	812	556	931	412	042	596	40
30	840	521	962	376	043	588	30
40	868	487	.2944	340	044	580	20
50	896	453	.3026	305	045	572	10
17° 00′	.2924	3.420	.3057	3.271	1.046	.9563	73° 00′
10	952	388	089	237	047	555	50
20	.2979	357	121	204	048	546	40
30	.3007	326	153	172	048	537	30
40	035	295	185	140	049	528	20
50	062	265	217	108	050	520	10
18° 00′	.3090	3.236	.3249	3.078	1.051	.9511	72° 00′
10	118	207	281	047	052	502	50
20	145	179	314	3.018	053	492	40
30	173	152	346	2.989	054	483	30
40	201	124	378	960	056	474	20
50	228	098	411	932	057	465	10
19° 00′	.3256	3.072	.3443	2.904	1.058	.9455	71° 00′
10	283	046	476	877	059	446	50
20	311	3.021	508	850	060	436	40
30	338	2.996	541	824	061	426	30
40	365	971	574	798	062	417	20
50	393	947	607	773	063	407	10
20° 00′	.3420	2.924	.3640	2.747	1.064	.9397	70° 00′
10	448	901	673	723	065	387	50
20	475	878	706	699	066	377	40
30	502	855	739	675	068	367	30
40	529	833	772	651	069	356	20
50	557	812	805	628	070	346	10
21° 00′	.3584	2.790	.3839	2.605	1.071	.9336	69° 00′
10	611	769	872	583	072	325	50
20	638	749	906	560	074	315	40
30	665	729	939	539	075	304	30
40	692	709	.3973	517	076	293	20
50	719	689	.4006	496	077	283	10
22° 00′	.3746	2.669	.4040	2.475	1.079	.9272	68° 00′
10	773	650	074	455	080	261	50
20	800	632	108	434	081	250	40
30	827	613	142	414	082	239	30
40	854	595	176	394	084	228	20
50	881	577	210	375	085	216	10
23° 00′	.3907	2.559	.4245	2.356	1.086	.9205	67° 00′
10	934	542	279	337	088	194	50
20	961	525	314	318	089	182	40
30	.3987	508	348	300	090	171	30
40	.4014	491	383	282	092	159	20
50	041	475	417	264	093	147	10
24° 00′	.4067	2.459	.4452	2.246	1.095	.9135	66° 00′
	cos θ	sec θ	cot θ	tan θ	csc θ	sin θ	Angle θ

Table II (continued)

Angle θ	sin θ	csc θ	tan θ	cot θ	sec θ	cos θ	
24° 00′	.4067	2.459	.4452	2.246	1.095	.9135	66° 00′
10	094	443	487	229	096	124	50
20	120	427	522	211	097	112	40
30	147	411	557	194	099	100	30
40	173	396	592	177	100	088	20
50	200	381	628	161	102	075	10
25° 00′	.4226	2.366	.4663	2.145	1.103	.9063	65° 00′
10	253	352	699	128	105	051	50
20	279	337	734	112	106	038	40
30	305	323	770	097	108	026	30
40	331	309	806	081	109	013	20
50	358	295	841	066	111	.9001	10
26° 00′	.4384	2.281	.4877	2.050	1.113	.8988	64° 00′
10	410	268	913	035	114	975	50
20	436	254	950	020	116	962	40
30	462	241	.4986	2.006	117	949	30
40	488	228	.5022	1.991	119	936	20
50	514	215	059	977	121	923	10
27° 00′	.4540	2.203	.5095	1.963	1.122	.8910	63° 00′
10	566	190	132	949	124	897	50
20	592	178	169	935	126	884	40
30	617	166	206	921	127	870	30
40	643	154	243	907	129	857	20
50	669	142	280	894	131	843	10
28° 00′	.4695	2.130	.5317	1.881	1.133	.8829	62° 00′
10	720	118	354	868	134	816	50
20	746	107	392	855	136	802	40
30	772	096	430	842	138	788	30
40	797	085	467	829	140	774	20
50	823	074	505	816	142	760	10
29° 00′	.4848	2.063	.5543	1.804	1.143	.8746	61° 00′
10	874	052	581	792	145	732	50
20	899	041	619	780	147	718	40
30	924	031	658	767	149	704	30
40	950	020	696	756	151	689	20
50	.4975	010	735	744	153	675	10
30° 00′	.5000	2.000	.5774	1.732	1.155	.8660	60° 00′
10	025	1.990	812	720	157	646	50
20	050	980	851	709	159	631	40
30	075	970	890	698	161	616	30
40	100	961	930	686	163	601	20
50	125	951	.5969	675	165	587	10
31° 00′	.5150	1.942	.6009	1.664	1.167	.8572	59° 00′
10	175	932	048	653	169	557	50
20	200	923	088	643	171	542	40
30	225	914	128	632	173	526	30
40	250	905	168	621	175	511	20
50	275	896	208	611	177	496	10
32° 00′	.5299	1.887	.6249	1.600	1.179	.8480	58° 00′
	cos θ	sec θ	cot θ	tan θ	csc θ	sin θ	Angle θ

Table II (continued)

Angle θ	sin θ	csc θ	tan θ	cot θ	sec θ	cos θ	
32° 00′	.5299	1.887	.6249	1.600	1.179	.8480	58° 00′
10	324	878	289	590	181	465	50
20	348	870	330	580	184	450	40
30	373	861	371	570	186	434	30
40	398	853	412	560	188	418	20
50	422	844	453	550	190	403	10
33° 00′	.5446	1.836	.6494	1.540	1.192	.8387	57° 00′
10	471	828	536	530	195	371	50
20	495	820	577	520	197	355	40
30	519	812	619	511	199	339	30
40	544	804	661	501	202	323	20
50	568	796	703	492	204	307	10
34° 00′	.5592	1.788	.6745	1.483	1.206	.8290	56° 00′
10	616	781	787	473	209	274	50
20	640	773	830	464	211	258	40
30	664	766	873	455	213	241	30
40	688	758	916	446	216	225	20
50	712	751	.6959	437	218	208	10
35° 00′	.5736	1.743	.7002	1.428	1.221	.8192	55° 00′
10	760	736	046	419	223	175	50
20	783	729	089	411	226	158	40
30	807	722	133	402	228	141	30
40	831	715	177	393	231	124	20
50	854	708	221	385	233	107	10
36° 00′	.5878	1.701	.7265	1.376	1.236	.8090	54° 00′
10	901	695	310	368	239	073	50
20	925	688	355	360	241	056	40
30	948	681	400	351	244	039	30
40	972	675	445	343	247	021	20
50	.5995	668	490	335	249	.8004	10
37° 00′	.6018	1.662	.7536	1.327	1.252	.7986	53° 00′
10	041	655	581	319	255	969	50
20	065	649	627	311	258	951	40
30	088	643	673	303	260	934	30
40	111	636	720	295	263	916	20
50	134	630	766	288	266	898	10
38° 00′	.6157	1.624	.7813	1.280	1.269	.7880	52° 00′
10	180	618	860	272	272	862	50
20	202	612	907	265	275	844	40
30	225	606	.7954	257	278	826	30
40	248	601	.8002	250	281	808	20
50	271	595	050	242	284	790	10
39° 00′	.6293	1.589	.8098	1.235	1.287	.7771	51° 00′
10	316	583	146	228	290	753	50
20	338	578	195	220	293	735	40
30	361	572	243	213	296	716	30
40	383	567	292	206	299	698	20
50	406	561	342	199	302	679	10
40° 00′	.6428	1.556	.8391	1.192	1.305	.7660	50° 00′
	cos θ	sec θ	cot θ	tan θ	csc θ	sin θ	Angle θ

Table II (continued)

Angle θ	sin θ	csc θ	tan θ	cot θ	sec θ	cos θ	
40° 00′	.6428	1.556	.8391	1.192	1.305	.7660	50° 00′
10	450	550	441	185	309	642	50
20	472	545	491	178	312	623	40
30	494	540	541	171	315	604	30
40	517	535	591	164	318	585	20
50	539	529	642	157	322	566	10
41° 00′	.6561	1.524	.8693	1.150	1.325	.7547	49° 00′
10	583	519	744	144	328	528	50
20	604	514	796	137	332	509	40
30	626	509	847	130	335	490	30
40	648	504	899	124	339	470	20
50	670	499	.8952	117	342	451	10
42° 00′	.6691	1.494	.9004	1.111	1.346	.7431	48° 00′
10	713	490	057	104	349	412	50
20	734	485	110	098	353	392	40
30	756	480	163	091	356	373	30
40	777	476	217	085	360	353	20
50	799	471	271	079	364	333	10
43° 00′	.6820	1.466	.9325	1.072	1.367	.7314	47° 00′
10	841	462	380	066	371	294	50
20	862	457	435	060	375	274	40
30	884	453	490	054	379	254	30
40	905	448	545	048	382	234	20
50	926	444	601	042	386	214	10
44° 00′	.6947	1.440	.9657	1.036	1.390	.7193	46° 00′
10	967	435	713	030	394	173	50
20	.6988	431	770	024	398	153	40
30	.7009	427	827	018	402	133	30
40	030	423	884	012	406	112	20
50	050	418	.9942	006	410	092	10
45° 00′	.7071	1.414	1.000	1.000	1.414	.7071	45° 00′
	cos θ	sec θ	cot θ	tan θ	csc θ	sin θ	Angle θ

Table III Values of Trigonometric Functions (Real Number x or θ Radians)

Real Number x or θ radians	$\sin x$ or $\sin\theta$	$\csc x$ or $\csc\theta$	$\tan x$ or $\tan\theta$	$\cot x$ or $\cot\theta$	$\sec x$ or $\sec\theta$	$\cos x$ or $\cos\theta$
0.00	0.0000	No value	0.0000	No value	1.000	1.000
.01	.0100	100.0	.0100	100.0	1.000	1.000
.02	.0200	50.00	.0200	49.99	1.000	0.9998
.03	.0300	33.34	.0300	33.32	1.000	0.9996
.04	.0400	25.01	.0400	24.99	1.001	0.9992
0.05	0.0500	20.01	0.0500	19.98	1.001	0.9988
.06	.0600	16.68	.0601	16.65	1.002	.9982
.07	.0699	14.30	.0701	14.26	1.002	.9976
.08	.0799	12.51	.0802	12.47	1.003	.9968
.09	.0899	11.13	.0902	11.08	1.004	.9960
0.10	0.0998	10.02	0.1003	9.967	1.005	0.9950
.11	.1098	9.109	.1104	9.054	1.006	.9940
.12	.1197	8.353	.1206	8.293	1.007	.9928
.13	.1296	7.714	.1307	7.649	1.009	.9916
.14	.1395	7.166	.1409	7.096	1.010	.9902
0.15	0.1494	6.692	0.1511	6.617	1.011	0.9888
.16	.1593	6.277	.1614	6.197	1.013	.9872
.17	.1692	5.911	.1717	5.826	1.015	.9856
.18	.1790	5.586	.1820	5.495	1.016	.9838
.19	.1889	5.295	.1923	5.200	1.018	.9820
0.20	0.1987	5.033	0.2027	4.933	1.020	0.9801
.21	.2085	4.797	.2131	4.692	1.022	.9780
.22	.2182	4.582	.2236	4.472	1.025	.9759
.23	.2280	4.386	.2341	4.271	1.027	.9737
.24	.2377	4.207	.2447	4.086	1.030	.9713
0.25	0.2474	4.042	0.2553	3.916	1.032	0.9689
.26	.2571	3.890	.2660	3.759	1.035	.9664
.27	.2667	3.749	.2768	3.613	1.038	.9638
.28	.2764	3.619	.2876	3.478	1.041	.9611
.29	.2860	3.497	.2984	3.351	1.044	.9582
0.30	0.2955	3.384	0.3093	3.233	1.047	0.9553
.31	.3051	3.278	.3203	3.122	1.050	.9523
.32	.3146	3.179	.3314	3.018	1.053	.9492
.33	.3240	3.086	.3425	2.920	1.057	.9460
.34	.3335	2.999	.3537	2.827	1.061	.9428
0.35	0.3429	2.916	0.3650	2.740	1.065	0.9394
.36	.3523	2.839	.3764	2.657	1.068	.9359
.37	.3616	2.765	.3879	2.578	1.073	.9323
.38	.3709	2.696	.3994	2.504	1.077	.9287
.39	.3802	2.630	.4111	2.433	1.081	.9249
0.40	0.3894	2.568	0.4228	2.365	1.086	0.9211
.41	.3986	2.509	.4346	2.301	1.090	.9171
.42	.4078	2.452	.4466	2.239	1.095	.9131
.43	.4169	2.399	.4586	2.180	1.100	.9090
.44	.4259	2.348	.4708	2.124	1.105	.9048
0.45	0.4350	2.299	0.4831	2.070	1.111	0.9004

Table III (continued)

Real Number x or θ radians	$\sin x$ or $\sin\theta$	$\csc x$ or $\csc\theta$	$\tan x$ or $\tan\theta$	$\cot x$ or $\cot\theta$	$\sec x$ or $\sec\theta$	$\cos x$ or $\cos\theta$
0.45	0.4350	2.299	0.4831	2.070	1.111	0.9004
.46	.4439	2.253	.4954	2.018	1.116	.8961
.47	.4529	2.208	.5080	1.969	1.122	.8916
.48	.4618	2.166	.5206	1.921	1.127	.8870
.49	.4706	2.125	.5334	1.875	1.133	.8823
0.50	0.4794	2.086	0.5463	1.830	1.139	0.8776
.51	.4882	2.048	.5594	1.788	1.146	.8727
.52	.4969	2.013	.5726	1.747	1.152	.8678
.53	.5055	1.978	.5859	1.707	1.159	.8628
.54	.5141	1.945	.5994	1.668	1.166	.8577
0.55	0.5227	1.913	0.6131	1.631	1.173	0.8525
.56	.5312	1.883	.6269	1.595	1.180	.8473
.57	.5396	1.853	.6410	1.560	1.188	.8419
.58	.5480	1.825	.6552	1.526	1.196	.8365
.59	.5564	1.797	.6696	1.494	1.203	.8309
0.60	0.5646	1.771	0.6841	1.462	1.212	0.8253
.61	.5729	1.746	.6989	1.431	1.220	.8196
.62	.5810	1.721	.7139	1.401	1.229	.8139
.63	.5891	1.697	.7291	1.372	1.238	.8080
.64	.5972	1.674	.7445	1.343	1.247	.8021
0.65	0.6052	1.652	0.7602	1.315	1.256	0.7961
.66	.6131	1.631	.7761	1.288	1.266	.7900
.67	.6210	1.610	.7923	1.262	1.276	.7838
.68	.6288	1.590	.8087	1.237	1.286	.7776
.69	.6365	1.571	.8253	1.212	1.297	.7712
0.70	0.6442	1.552	0.8423	1.187	1.307	0.7648
.71	.6518	1.534	.8595	1.163	1.319	.7584
.72	.6594	1.517	.8771	1.140	1.330	.7518
.73	.6669	1.500	.8949	1.117	1.342	.7452
.74	.6743	1.483	.9131	1.095	1.354	.7385
0.75	0.6816	1.467	0.9316	1.073	1.367	0.7317
.76	.6889	1.452	.9505	1.052	1.380	.7248
.77	.6961	1.436	.9697	1.031	1.393	.7179
.78	.7033	1.422	.9893	1.011	1.407	.7109
.79	.7104	1.408	1.009	.9908	1.421	.7038
0.80	0.7174	1.394	1.030	0.9712	1.435	0.6967
.81	.7243	1.381	1.050	.9520	1.450	.6895
.82	.7311	1.368	1.072	.9331	1.466	.6822
.83	.7379	1.355	1.093	.9146	1.482	.6749
.84	.7446	1.343	1.116	.8964	1.498	.6675
0.85	0.7513	1.331	1.138	0.8785	1.515	0.6600
.86	.7578	1.320	1.162	.8609	1.533	.6524
.87	.7643	1.308	1.185	.8437	1.551	.6448
.88	.7707	1.297	1.210	.8267	1.569	.6372
.89	.7771	1.287	1.235	.8100	1.589	.6294
0.90	0.7833	1.277	1.260	0.7936	1.609	0.6216
.91	.7895	1.267	1.286	.7774	1.629	.6137
.92	.7956	1.257	1.313	.7615	1.651	.6058
.93	.8016	1.247	1.341	.7458	1.673	.5978
.94	.8076	1.238	1.369	.7303	1.696	.5898
0.95	0.8134	1.229	1.398	0.7151	1.719	0.5817

Table III (continued)

Real Number x or θ radians	$\sin x$ or $\sin \theta$	$\csc x$ or $\csc \theta$	$\tan x$ or $\tan \theta$	$\cot x$ or $\cot \theta$	$\sec x$ or $\sec \theta$	$\cos x$ or $\cos \theta$
0.95	0.8134	1.229	1.398	0.7151	1.719	0.5817
.96	.8192	1.221	1.428	.7001	1.744	.5735
.97	.8249	1.212	1.459	.6853	1.769	.5653
.98	.8305	1.204	1.491	.6707	1.795	.5570
.99	.8360	1.196	1.524	.6563	1.823	.5487
1.00	0.8415	1.188	1.557	0.6421	1.851	0.5403
1.01	.8468	1.181	1.592	.6281	1.880	.5319
1.02	.8521	1.174	1.628	.6142	1.911	.5234
1.03	.8573	1.166	1.665	.6005	1.942	.5148
1.04	.8624	1.160	1.704	.5870	1.975	.5062
1.05	0.8674	1.153	1.743	0.5736	2.010	0.4976
1.06	.8724	1.146	1.784	.5604	2.046	.4889
1.07	.8772	1.140	1.827	.5473	2.083	.4801
1.08	.8820	1.134	1.871	.5344	2.122	.4713
1.09	.8866	1.128	1.917	.5216	2.162	.4625
1.10	0.8912	1.122	1.965	0.5090	2.205	0.4536
1.11	.8957	1.116	2.014	.4964	2.249	.4447
1.12	.9001	1.111	2.066	.4840	2.295	.4357
1.13	.9044	1.106	2.120	.4718	2.344	.4267
1.14	.9086	1.101	2.176	.4596	2.395	.4176
1.15	0.9128	1.096	2.234	0.4475	2.448	0.4085
1.16	.9168	1.091	2.296	.4356	2.504	.3993
1.17	.9208	1.086	2.360	.4237	2.563	.3902
1.18	.9246	1.082	2.427	.4120	2.625	.3809
1.19	.9284	1.077	2.498	.4003	2.691	.3717
1.20	0.9320	1.073	2.572	0.3888	2.760	0.3624
1.21	.9356	1.069	2.650	.3773	2.833	.3530
1.22	.9391	1.065	2.733	.3659	2.910	.3436
1.23	.9425	1.061	2.820	.3546	2.992	.3342
1.24	.9458	1.057	2.912	.3434	3.079	.3248
1.25	0.9490	1.054	3.010	0.3323	3.171	0.3153
1.26	.9521	1.050	3.113	.3212	3.270	.3058
1.27	.9551	1.047	3.224	.3102	3.375	.2963
1.28	.9580	1.044	3.341	.2993	3.488	.2867
1.29	.9608	1.041	3.467	.2884	3.609	.2771
1.30	0.9636	1.038	3.602	0.2776	3.738	0.2675
1.31	.9662	1.035	3.747	.2669	3.878	.2579
1.32	.9687	1.032	3.903	.2562	4.029	.2482
1.33	.9711	1.030	4.072	.2456	4.193	.2385
1.34	.9735	1.027	4.256	.2350	4.372	.2288
1.35	0.9757	1.025	4.455	0.2245	4.566	0.2190
1.36	.9779	1.023	4.673	.2140	4.779	.2092
1.37	.9799	1.021	4.913	.2035	5.014	.1994
1.38	.9819	1.018	5.177	.1931	5.273	.1896
1.39	.9837	1.017	5.471	.1828	5.561	.1798
1.40	0.9854	1.015	5.798	0.1725	5.883	0.1700
1.41	.9871	1.013	6.165	.1622	6.246	.1601
1.42	.9887	1.011	6.581	.1519	6.657	.1502
1.43	.9901	1.010	7.055	.1417	7.126	.1403
1.44	.9915	1.009	7.602	.1315	7.667	.1304
1.45	0.9927	1.007	8.238	0.1214	8.299	0.1205

Table III (continued)

Real Number x or θ radians	$\sin x$ or $\sin \theta$	$\csc x$ or $\csc \theta$	$\tan x$ or $\tan \theta$	$\cot x$ or $\cot \theta$	$\sec x$ or $\sec \theta$	$\cos x$ or $\cos \theta$
1.45	0.9927	1.007	8.238	0.1214	8.299	0.1205
1.46	.9939	1.006	8.989	.1113	9.044	.1106
1.47	.9949	1.005	9.887	.1011	9.938	.1006
1.48	.9959	1.004	10.98	.0910	11.03	.0907
1.49	.9967	1.003	12.35	.0810	12.39	.0807
1.50	0.9975	1.003	14.10	0.0709	14.14	0.0707
1.51	.9982	1.002	16.43	.0609	16.46	.0608
1.52	.9987	1.001	19.67	.0508	19.69	.0508
1.53	.9992	1.001	24.50	.0408	24.52	.0408
1.54	.9995	1.000	32.46	.0308	32.48	.0308
1.55	0.9998	1.000	48.08	0.0208	48.09	0.0208
1.56	.9999	1.000	92.62	.0108	92.63	.0108
1.57	1.000	1.000	1256	.0008	1256	.0008

SOLUTION KEY (Odd-Numbered Exercises)

[1.1]

1. $\tan\frac{\pi}{3} = \tan\frac{\pi^R}{3} = \sqrt{3}$ **3.** $\sec\frac{\pi}{6} = \sec\frac{\pi^R}{6} = \frac{2}{\sqrt{3}}$ **5.** $\cos 0 = \cos 0^R = 1$

7. By definition, $\tilde{x} = \pi - x = \frac{\pi}{3}$. Hence, by Property 3,

$$\cos\frac{2\pi}{3} = -\cos\frac{\pi}{3}$$

where the negative sign is chosen because the arc with length $\frac{2\pi}{3}$ terminates in Quadrant II. Since $\cos\frac{\pi}{3} = \cos\frac{\pi^R}{3} = \frac{1}{2}$, then

$$\cos\frac{2\pi}{3} = -\cos\frac{\pi^R}{3} = -\frac{1}{2}$$

9. By definition, $\tilde{x} = 2\pi - x = \frac{\pi}{4}$. Hence, by Property 3,

$$\cot\frac{7\pi}{4} = -\cot\frac{\pi}{4}$$

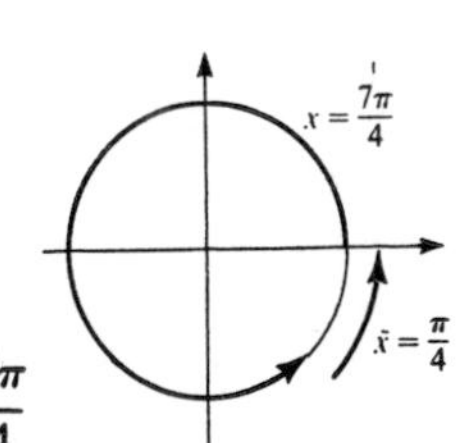

where the negative sign is chosen because the arc with length $\frac{7\pi}{4}$ terminates in Quadrant IV. Since $\cot\frac{\pi}{4} = \cot\frac{\pi^R}{4} = 1$, then

$$\cot\frac{7\pi}{4} = -\cot\frac{\pi^R}{4} = -1$$

11. $\cos \frac{5\pi}{4} = -\cos \frac{\pi}{4} = -\cos \frac{\pi^R}{4} = -\frac{1}{\sqrt{2}}$

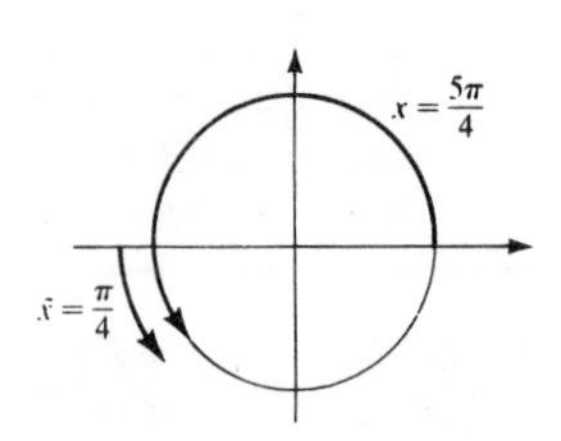

13. From Table III,

$$\cos 0.43 \approx 0.9090$$

15. By Property 3, $\tan 1.93 \approx -\tan 1.21$, where the negative sign is chosen because the arc with length 1.93 terminates in Quadrant II. Since, from Table III, $\tan 1.21 \approx 2.650$, then

$$\tan 1.93 \approx -\tan 1.21 \approx -2.650$$

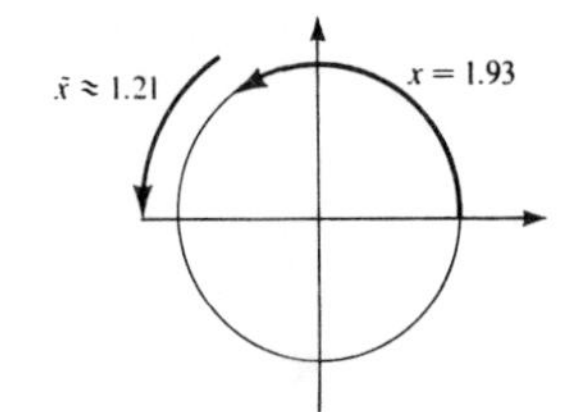

17. By Property 3, $\sin 5.83 \approx -\sin 0.45$, where the negative sign is chosen because the arc with length 5.83 terminates in Quadrant IV. Since, from Table III, $\sin 0.45 \approx 0.4350$, then

$$\sin 5.83 \approx -\sin 0.45 \approx -0.4350$$

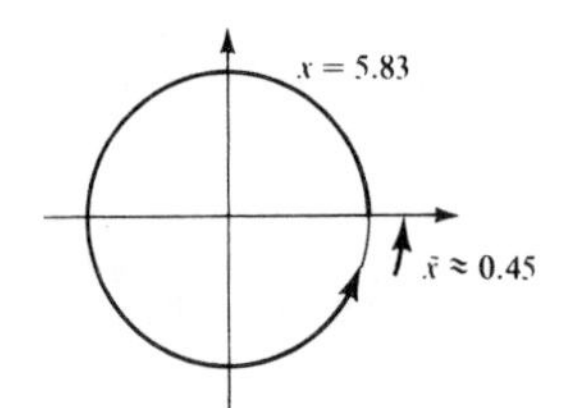

19. By Property 3, $\cos 1.92 \approx -\cos 1.22$. Since, from Table III, $\cos 1.22 \approx 0.3436$, then

$$\cos 1.92 \approx -\cos 1.22 \approx -0.3436$$

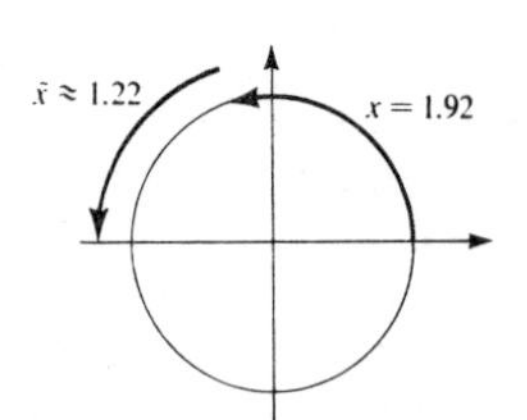

21. $\bar{x} \approx 0.86$. By Property 3, $\tan(-5.42) \approx \tan 0.86$. Since, from Table III, we have $\tan 0.86 \approx 1.162$, then

$$\tan(-5.42) \approx \tan 0.86 \approx 1.162$$

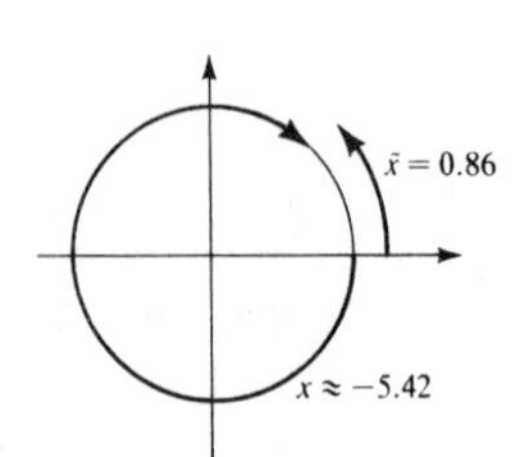

23. $\bar{x} \approx 0.78$. By Property 3, $\cos(-3.92) \approx -\cos 0.78$. Since, from Table III, $\cos 0.78 \approx 0.7109$, then

$$\cos(-3.92) = -\cos 0.78 \approx -0.7109$$

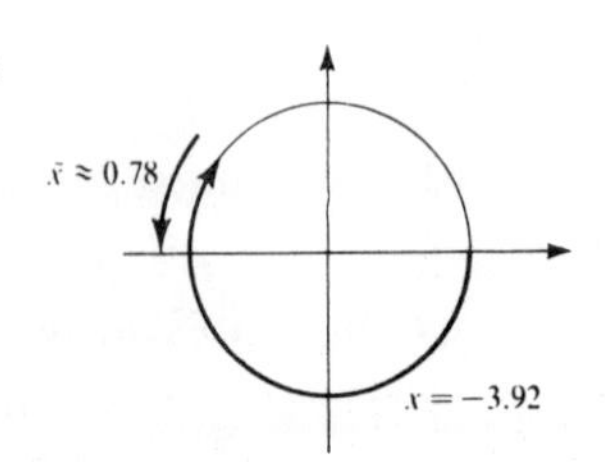

25. Since the cosine function is periodic with period $2\pi \approx 6.28$, then $\cos 7.21 \approx \cos(7.21 - 6.28) = \cos 0.93$. Since the arc with length 7.21 terminates in Quadrant I, then cos 7.21 is positive. Furthermore, from Table III we have $\cos 0.93 \approx 0.5978$. Hence

$$\cos 7.21 = \cos 0.93 \approx 0.5978$$

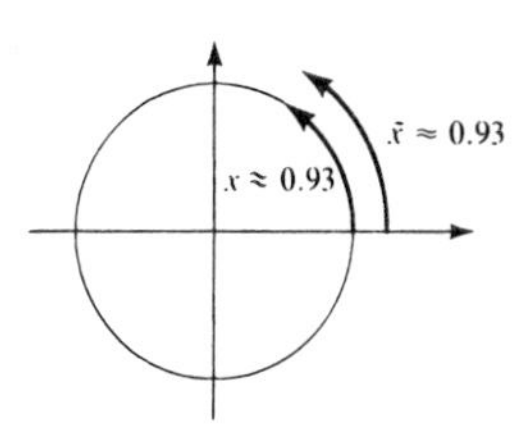

27. $\sin 11.30 \approx \sin(11.30 - 6.28)$
$= \sin 5.02 \approx -\sin 1.26$
≈ -0.9521

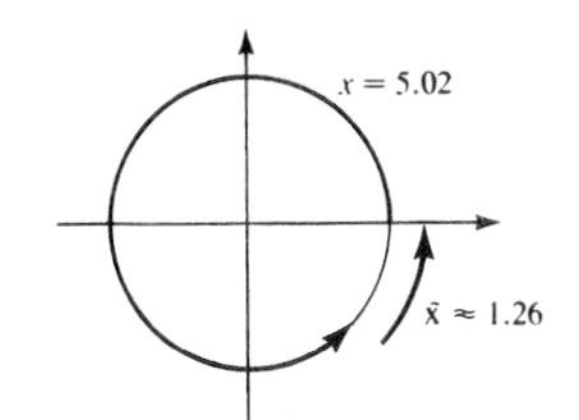

29. $\sin 15.60 = \sin(15.60 - 2 \cdot 6.28)$
$\approx \sin 3.04 \approx \sin 0.10$
≈ 0.0998

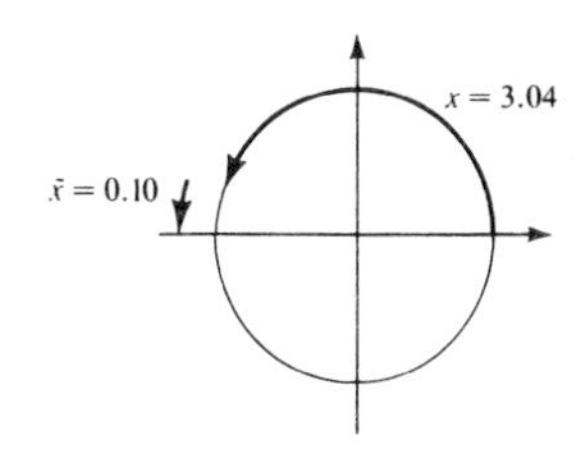

31. In $i = I_{max} \sin \omega t$, we substitute 0.06 for I_{max}, 350 for ω, and 0.02 for t and obtain,

$$i = 0.06 \sin 350(0.02) = 0.06 \sin 7$$

Since $\sin 7 = \sin(7 - 2\pi) \approx \sin 0.72$, we have

$$i \approx 0.06 \sin 0.72 \approx 0.06(0.6594) \approx 0.040$$

Hence, i is 0.040 ampere to the nearest thousandth.

33. $i = 0.06 \sin[350(0.1)]$
$= 0.06 \sin 35$
$= 0.06 \sin(35 - 5 \cdot 2\pi)$
$= 0.06 \sin 3.60$
$\approx 0.06(-\sin 0.46)$
$\approx 0.06(-0.4439) \approx -0.027$

Hence, i is -0.027 ampere to the nearest thousandth.

35. $i = 0.06 \sin[350(1.0)]$
$= 0.06 \sin 350$
$= 0.06 \sin(350 - 55 \cdot 2\pi)$
$= 0.06 \sin 4.60 \approx 0.06(-\sin 1.46)$
$\approx 0.06(-0.9939) \approx -0.060$

Hence, i is -0.060 ampere to the nearest thousandth.

37. In $d = k \cos \omega t$, we substitute -8 for k, 5 for ω, and 2 for t and obtain

$d = -8 \cos(5)(2) = -8 \cos 10$
$\approx -8 \cos(10 - 6.28) = -8 \cos 3.72$
$\approx -8(-\cos 0.58) \approx -8(-0.8365)$
≈ 6.7

Hence, d is 6.7 centimeters to the nearest tenth, below rest position. When t is 5, we have

$d = -8 \cos[5(5)] = -8 \cos 25$
$\approx -8 \cos(25 - 3 \cdot 6.28)$
$= -8 \cos 6.16 = -8 \cos 0.12$
$\approx -8(0.9928) \approx -7.9$

Hence, d is 7.9 centimeters to the nearest tenth, below rest position.

39. In the formula $d = k \cos \omega t$, we substitute -6 for k, 5 for ω, and 0 for t and obtain

$d = -6 \cos(5)(0) = -6 \cos 0$
$= -6(1) = -6$

Since the sine function has period 2π, for $t = 0$, 2π, and 4π, d is 6 centimeters below rest position.

41. In the formula $d = k \cos \omega t$, we substitute 15 for d, 5 for ω, and 1 for t, and obtain

$$15 = k \cos (5)(1)$$

$$k = \frac{15}{\cos 5} \approx \frac{15}{\cos 1.28} \approx \frac{15}{0.2867}$$

$$\approx 52.3$$

Hence k is 52.3 centimeters to the nearest tenth.

43. Substitute 2.2 for t and 18 for A and obtain

$$\begin{aligned} d &= 18 \sin 2\pi(2.2) \\ &\approx 18 \sin 13.82 \\ &\approx 18 \sin (13.82 - 2 \cdot 6.28) \approx 18 \sin 1.26 \\ &= 18(0.9521) \approx 17.1 \end{aligned}$$

Hence d is 17.1 centimeters to the nearest tenth.

45. Substitute 22 for d and $\frac{1}{4}$ for t and obtain

$$22 = A \sin 2\pi \left(\frac{1}{4}\right) = A \sin \frac{\pi}{2}$$

$$A = \frac{22}{\sin (\pi/2)} = \frac{22}{1} = 22$$

Hence A is 22 centimeters.

[1.2]

1. It may be helpful to sketch $y = \sin x$, $0 \le x \le 2\pi$, as a reference (dashed curve). Then, since the amplitude of $y = 2 \sin x$ is 2, we can sketch the desired graph on the same coordinate system over the interval $0 \le x \le 2\pi$ by making each ordinate 2 times the corresponding ordinate of the graph of $y = \sin x$. We can then extend this cycle to include the entire interval $-2\pi \le x \le 2\pi$, as shown in the figure. For $y = 2 \sin x$, the fundamental period p is the same as for $y = \sin x$, namely, $p = 2\pi$.

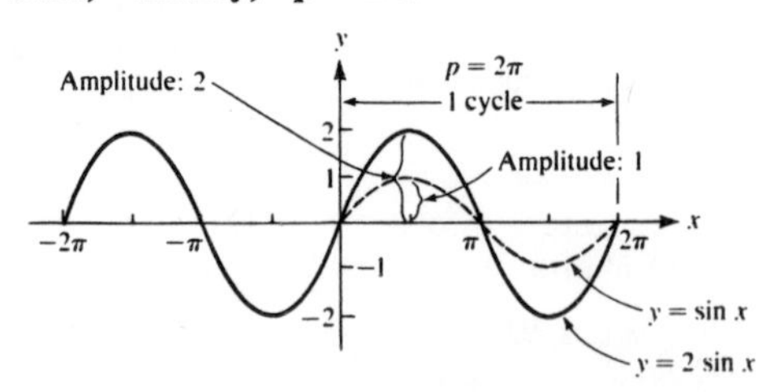

3. Sketch $y = \cos x$, $0 \le x \le 2\pi$, as a reference (dashed curve). Then, since the amplitude of $y = \frac{1}{2} \cos x$ is $\frac{1}{2}$, we can sketch the desired graph on the same coordinate system over the interval $0 \le x \le 2\pi$ by making each ordinate one-half of the corresponding ordinate of the graph of $y = \cos x$. We can then extend this cycle to include the entire interval $-2\pi \le x \le 2\pi$, as shown in the figure. For $y = \frac{1}{2} \cos x$, the fundamental period p is the same as for $y = \cos x$, namely, $p = 2\pi$.

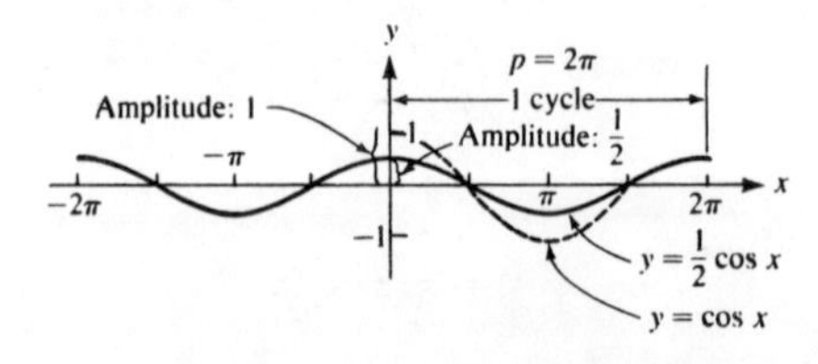

5. Sketch $y = \sin x$, $0 \le x \le 2\pi$, as a reference (dashed curve). Then, since the amplitude of $y = -4 \sin x$ is $|-4|$, we can sketch the desired graph on the same coordinate system over the interval $0 \le x \le 2\pi$ by making each ordinate -4 times the corresponding ordinate of the graph $y = \sin x$. We can then extend this cycle to include the entire interval $-2\pi \le x \le 2\pi$, as shown in the figure. For $y = -4 \sin x$, the fundamental period p is the same as for $y = \sin x$, namely, $p = 2\pi$.

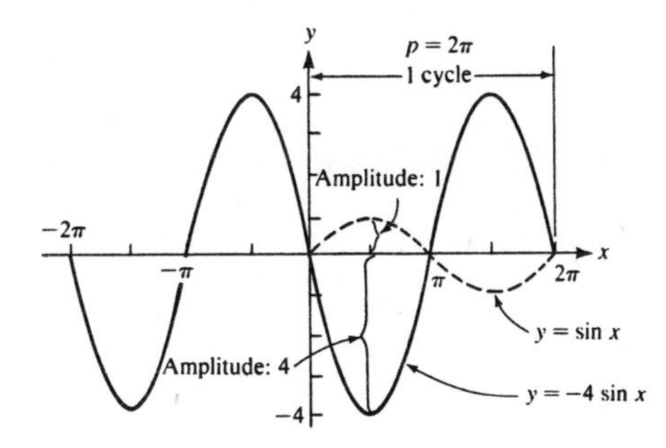

7. First sketch a cycle of $y = \sin x$, $0 \le x \le 2\pi$, as a reference (dashed curve). Since, by Property 3,

$$p = \frac{2\pi}{|B|} = \frac{2\pi}{2} = \pi$$

we next sketch a cycle of the graph of $y = \sin 2x$ over the interval $0 \le x \le \pi$ on the same coordinate system and extend the cycle obtained over the interval $-2\pi \le x \le 2\pi$.

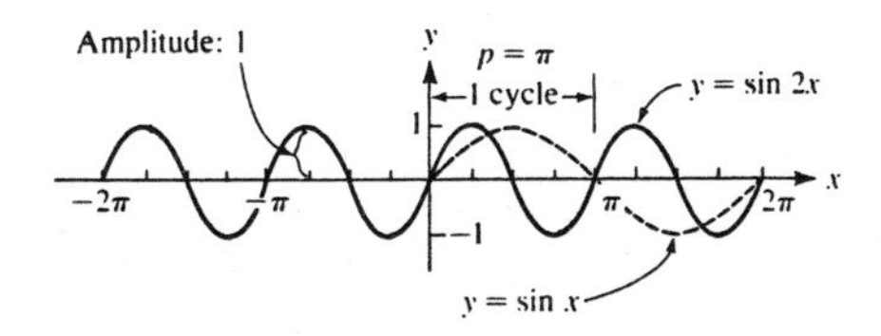

9. Sketch a cycle of $y = \cos x$, $0 \le x \le 2\pi$, as a reference (dashed curve). Since

$$p = \frac{2\pi}{|B|} = \frac{2\pi}{1/3} = 6\pi$$

we next sketch part of the graph of $y = \cos \frac{1}{3}x$ over the interval $0 \le x \le 3\pi/2$ on the same coordinate system and extend the graph obtained over the interval $-2\pi \le x \le 2\pi$.

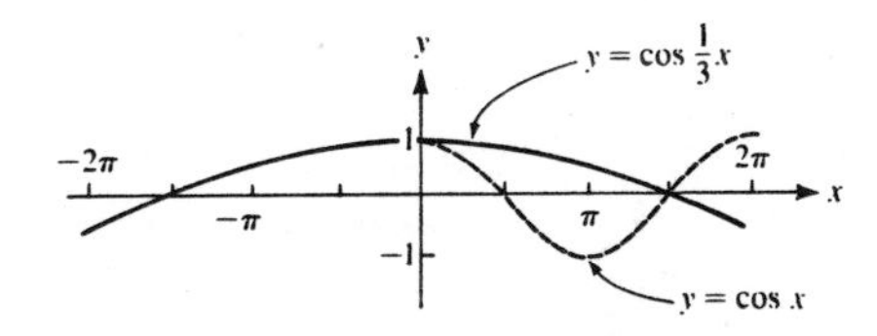

11. Sketch the graph of $y = \cos x$ as a reference (dashed curve). Each ordinate of the graph of $y = -2 \cos x$ is the negative of the ordinate of the graph of $y = 2 \cos x$. Since, by Property 3,

$$p = \frac{2\pi}{|B|} = 2\pi$$

there is one cycle in the interval $0 \le x \le 2\pi$.

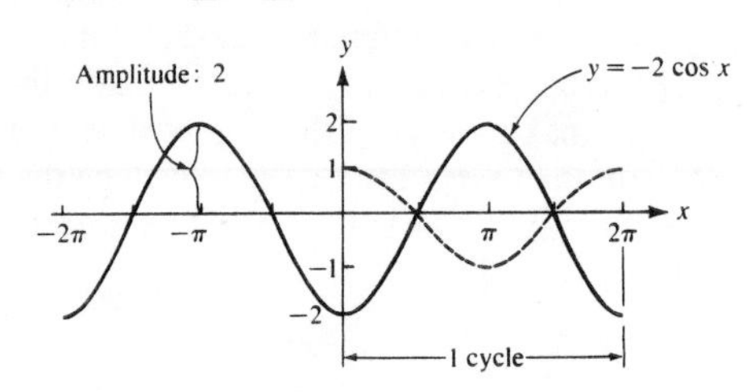

13. Sketch the graph of $y = \sin x$ as a reference (dashed curve). Each ordinate of the graph of $y = -3 \sin 2x$ is the negative of the ordinate of the graph of $y = 3 \sin 2x$. Since

$$p = \frac{2\pi}{|B|} = \frac{2\pi}{2} = \pi$$

there is one cycle in the interval $0 \le x \le \pi$.

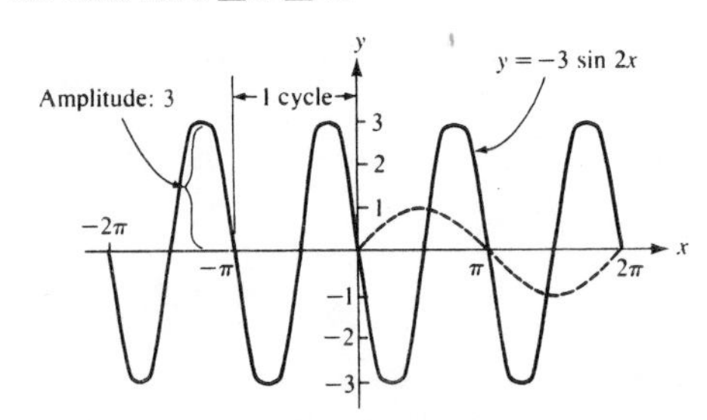

15. From Property 3, note the following things about this sine wave:

1. It has amplitude 1.
2. It has period $2\pi/1 = 2\pi$.
3. It is π units to the left of the graph of $y = \sin x$.

First sketch the graph of $y = \sin x$ (dashed curve) and then with the above facts, sketch the graph shown.

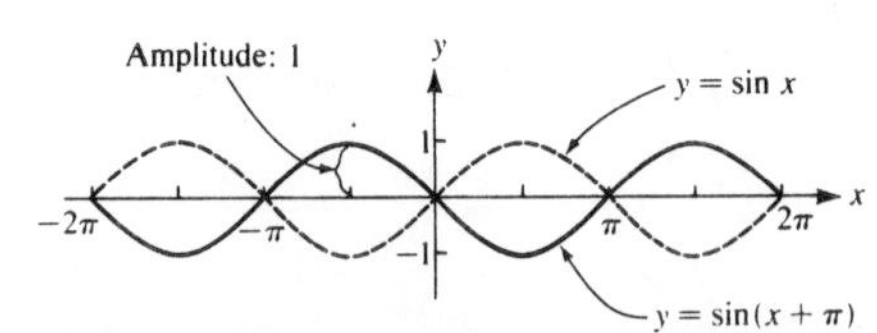

17. From Property 3, note the following things about this cosine wave:

1. It has amplitude 2.
2. It has period $2\pi/1 = 2\pi$.
3. It is $\pi/2$ units to the right of the graph of $y = 2 \cos x$.

First sketch the graph of $y = 2 \cos x$ and then with the above facts, sketch the graph shown.

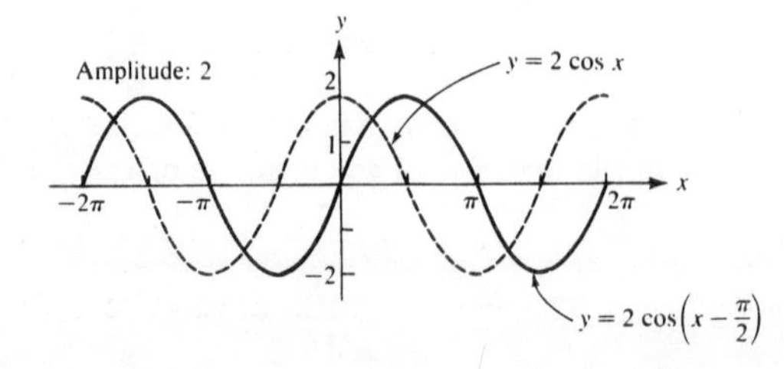

19. From Property 3, note the following things about this sine wave:

1. It has amplitude 3.
2. It has a period $2\pi/2 = \pi$.
3. It is $\pi/3$ units to the right of the graph of $y = 3 \sin 2x$.

First sketch the graph $y = 3 \sin 2x$ and then with the above facts, sketch the graph shown.

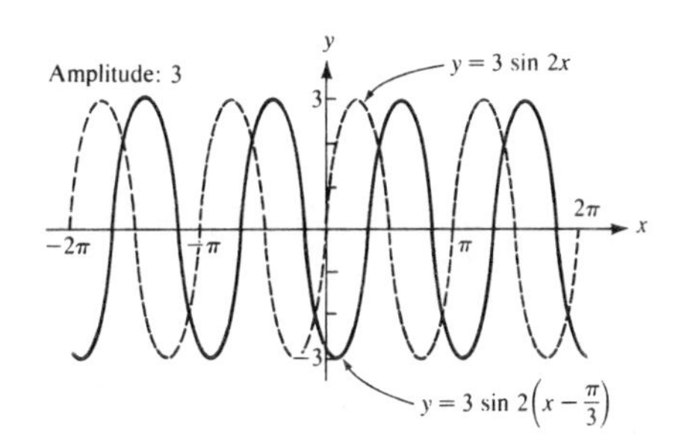

21. By Property 3, note the following properties of the graph:

1. It is a sine wave.
2. It has amplitude 2.
3. It has period $2\pi/\pi = 2$.

Since the period is 2, use integers as elements of the domain. Scaling the x axis in integral units facilitates sketching the graph.

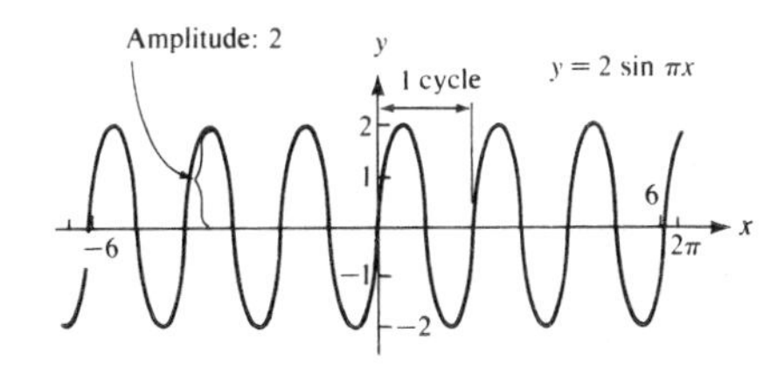

23. By Property 3, note the following properties of the graph:

1. It is a cosine wave.
2. It has amplitude $|-1/2| = 1/2$.
3. It has period $\dfrac{2\pi}{\pi/3} = 6$.

Since the period is 6, use integers as elements of the domain. Scaling the x axis in integral units facilitates sketching the graph.

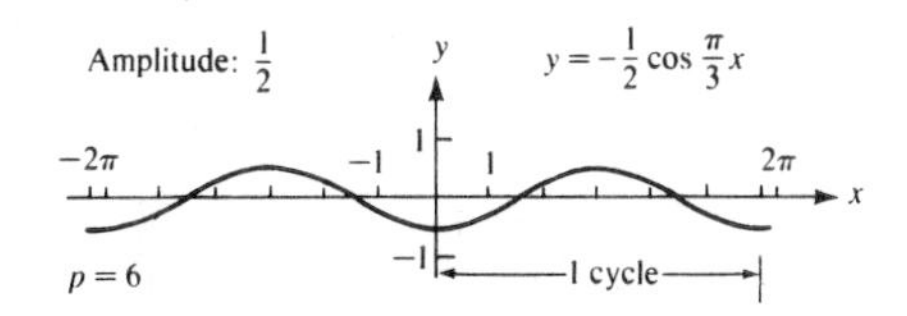

25. See the graph of $y = \sin 2x$ on page 79 (Exercise 7). The zeros are the x intercepts $-2\pi, -\dfrac{3\pi}{2}, -\pi, -\dfrac{\pi}{2}, 0, \dfrac{\pi}{2}, \pi, \dfrac{3\pi}{2}, 2\pi$ over the interval $-2\pi \leq x \leq 2\pi$.

27. See the graph of $y = \cos\frac{1}{3}x$ on page 79 (Exercise 9). The zeros are the x intercepts $-\frac{3\pi}{2}, \frac{3\pi}{2}$ over the interval $-2\pi \le x \le 2\pi$.

29. See the graph of $y = 2\sin\pi x$ on page 81 (Exercise 21). The zeros are the x intercepts $-6, -5, -4, -3, -2, -1, 0, 1, 2, 3, 4, 5, 6$ over the interval $-2\pi \le x \le 2\pi$.

[1.3]

1. Since csc $2x$ is undefined for $2x = k\pi$, $k \in J$, by Property 1 there are vertical asymptotes at $x = -2\pi, -\frac{3\pi}{2}, -\pi, -\frac{\pi}{2}, 0, \frac{\pi}{2}, \pi, \frac{3\pi}{2}, 2\pi$. By Property 2, the period p is given by

$$p = \frac{2\pi}{|2|} = \pi$$

Thus the graph of $y = \csc 2x$ is similar to that of $y = \csc x$, except each cycle is "contracted" into an interval of π units instead of 2π. The graph is sketched below.

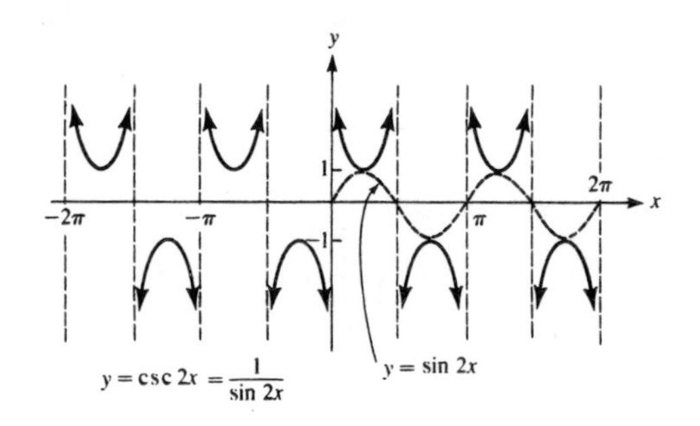

Note: For each x, $\csc 2x = \frac{1}{\sin 2x}$, $x \neq k\cdot\frac{\pi}{2}$, $k \in J$.

3. Since cot $2x$ is undefined for $2x = k\pi$, $k \in J$, by Property 1 there are vertical asymptotes at $x = -2\pi, -\frac{3\pi}{2}, -\pi, -\frac{\pi}{2}, 0, \frac{\pi}{2}, \pi, \frac{3\pi}{2}, 2\pi$. By Property 2, the period p is given by $p = \frac{\pi}{2}$. Thus the graph of $y = \cot 2x$ is similar to that of $y = \cot x$, except each cycle is "contracted" into an interval of $\frac{\pi}{2}$ units instead of π. The graph is sketched below.

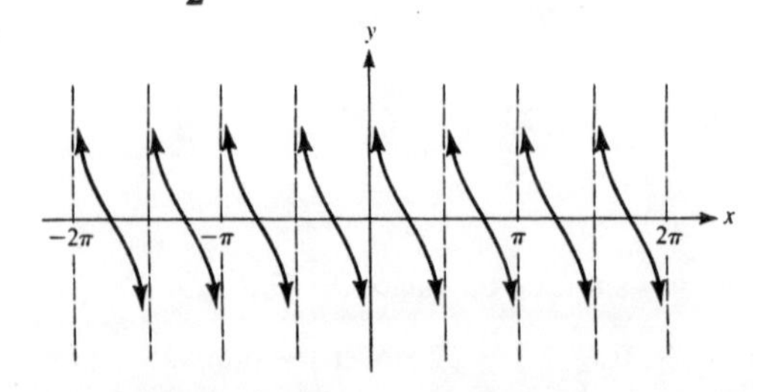

5. Since $y = \tan \frac{1}{2}x$ is undefined for $\frac{1}{2}x = \frac{\pi}{2} + k\pi$, $k \in J$, there are vertical asymptotes at $x = -\pi$ and π. The period p is given by $p = \frac{\pi}{\frac{1}{2}} = 2\pi$. Thus the graph of $y = \tan \frac{1}{2}x$ is similar to that of $y = \tan x$ except that each cycle is "spread out" over an interval of 2π units instead of π. The graph is sketched below.

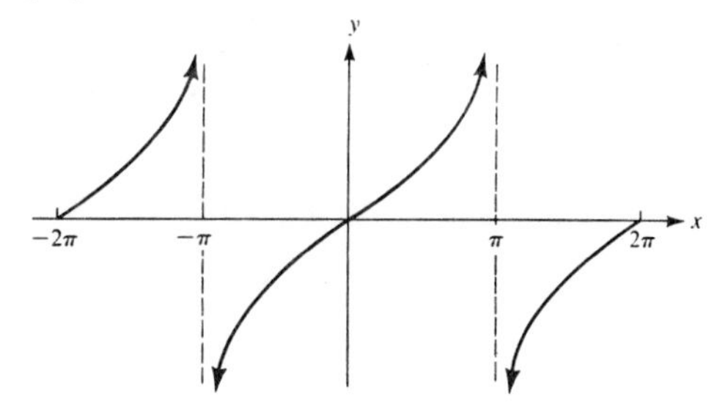

7. Since $2 \sec \frac{1}{2}x$ is undefined for $\frac{1}{2}x = \frac{\pi}{2} + k\pi$, $k \in J$, there are vertical asymptotes at $x = -\pi$ and π. The period p is given by $p = \frac{2\pi}{\frac{1}{2}} = 4\pi$. Thus the graph of $y = \sec \frac{1}{2}x$ is similar to that of $y = \sec x$ except that each cycle is "spread out" over an interval of 4π units instead of 2π. Also, since $2 \sec \frac{1}{2}x$ is double $\sec \frac{1}{2}x$, for each x, ordinates of $y = 2 \sec \frac{1}{2}x$ are double that of $y = \sec \frac{1}{2}x$. The graph is sketched below.

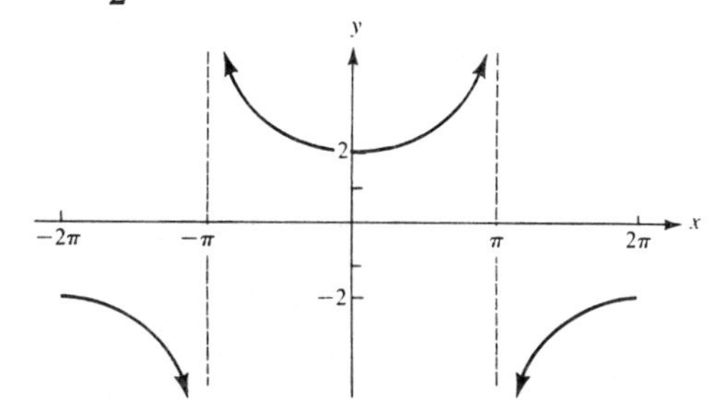

9. Comparing $y = \tan\left(x + \frac{\pi}{3}\right)$ with $y = \tan x$ (see Property 3), we see that $\frac{\pi}{3}$ is the phase shift, and the graph of $y = \tan\left(x + \frac{\pi}{3}\right)$ is $\frac{\pi}{3}$ units to the left of the graph of $y = \tan x$. First we sketch the graph of $y = \tan x$ and then shift it $\frac{\pi}{3}$ units to the left.

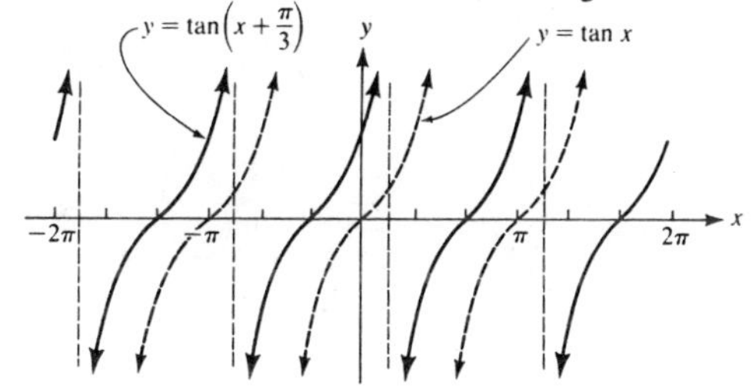

11. Comparing $y = \sec\left(x - \frac{\pi}{4}\right)$ with $y = \sec x$ (see Property 3), we see that $\frac{\pi}{4}$ is the phase shift, and the graph of $y = \sec\left(x - \frac{\pi}{4}\right)$ is $\frac{\pi}{4}$ units to the right of the graph of $y = \sec x$. First we sketch the graph of $y = \sec x$ and then shift it $\frac{\pi}{4}$ units to the right.

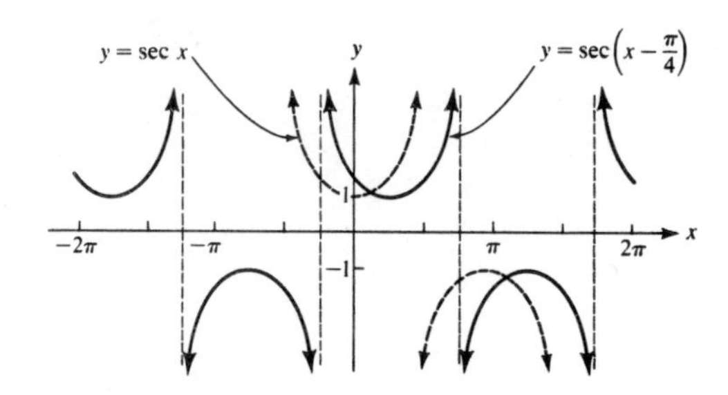

[1.4]

1. First sketch the graphs of $y = \sin x$ and $y = \cos x$ (dashed curves) on the same coordinate system over $0 \leq x \leq 2\pi$. The ordinate of the graph of $y = \sin x + \cos x$ at each point x on the x axis is the algebraic sum of the corresponding ordinates of $y = \sin x$ and $y = \cos x$. The ordinate can be approximated graphically by "adding" the directed line segments from the x axis to the curves from each point x. Then, duplicate this cycle of the curve over $-2\pi \leq x \leq 0$.

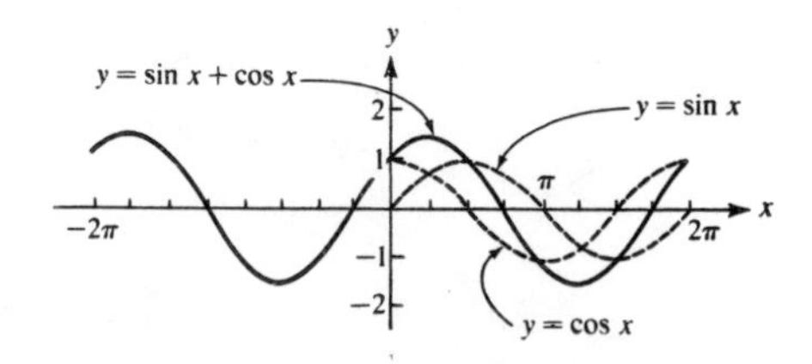

3. First sketch the graphs of $y = \sin 2x$ and $y = \frac{1}{2}\cos x$ (dashed curves) on the same coordinate system over $0 \leq x \leq 2\pi$. Graph $y = \sin 2x + \frac{1}{2}\cos x$ by "adding" ordinates. Duplicate this cycle of the curve over $-2\pi \leq x \leq 0$.

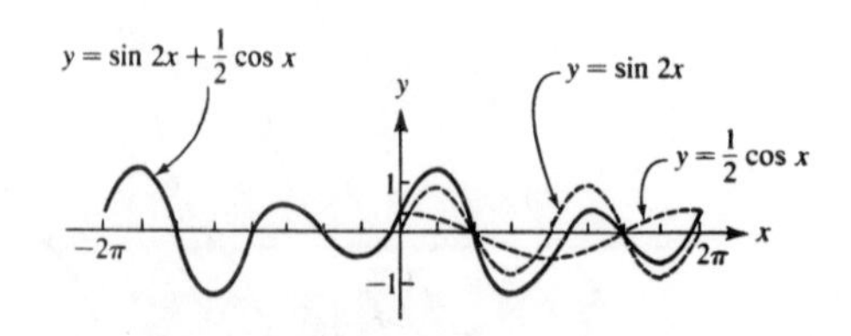

5. First sketch the graphs of $y = \sin x$ and $y = -2\cos x$ (dashed curves) on the same coordinate system over $0 \leq x \leq 2\pi$. Graph $y = \sin x - 2\cos x$ by "adding" the ordinates of $y = -2\cos x$ to the ordinates of $y = \sin x$ for each x. Then duplicate this cycle of the curve over the interval $-2\pi \leq x \leq 0$.

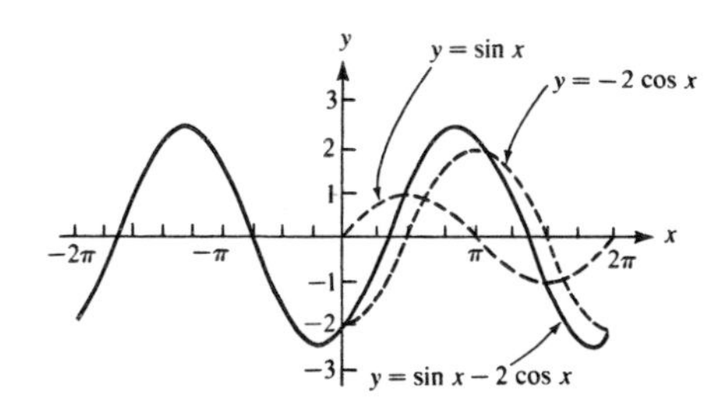

7 First sketch the graphs of $y=3 \sin x$ and $y=1$ on the same coordinate system over $0 \le x \le 2\pi$. Graph $y=3 \sin x+1$ by "adding" ordinates. Then duplicate this cycle over $-2\pi \le x \le 0$.

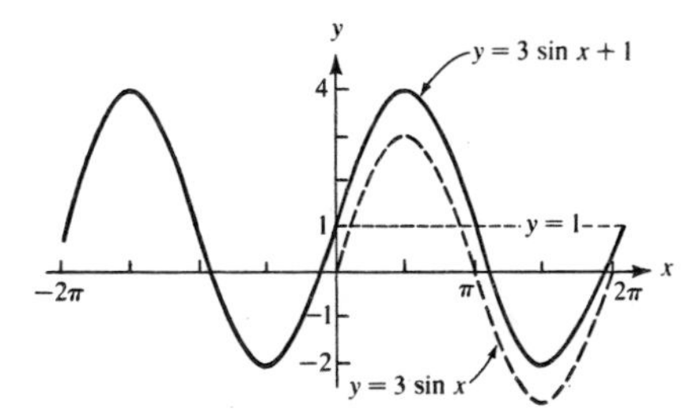

9. First sketch the graphs of $y=x$ and $y=\cos x$ on the same coordinate system over the given interval. Then use the method of "adding" of ordinates.

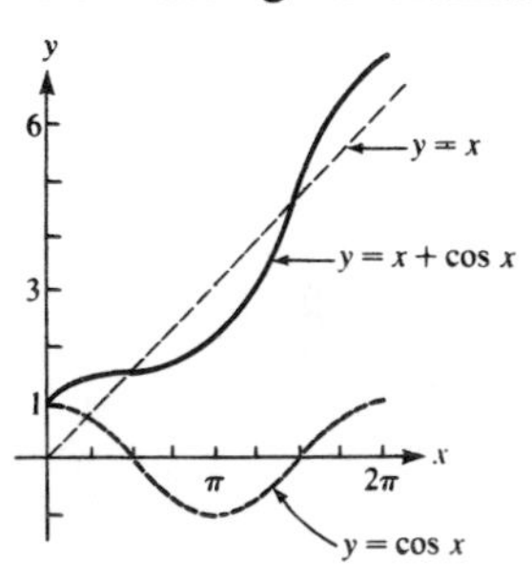

This function, while not periodic, has a graph over the interval $-2\pi \le x \le 0$, which "wraps" around $y=x$ as shown over $0 \le x \le 2\pi$.

11. First sketch the graphs of $y=x$ and $y=-\sin x$ on the same coordinate system over the given interval. Then use the method of "adding" ordinates.

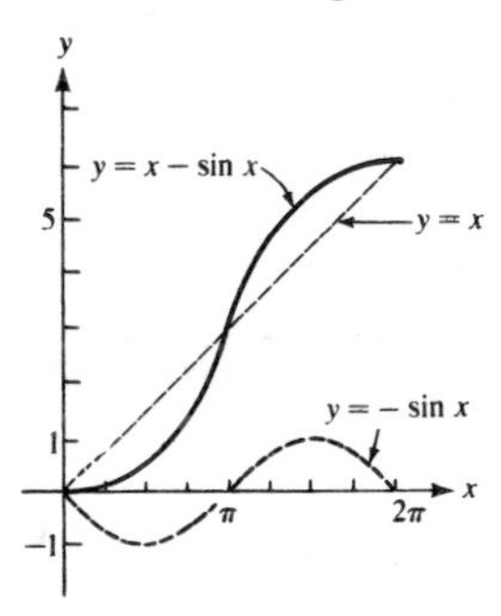

This function, while not periodic, has a graph over the interval $-2\pi \le x \le 0$, which "wraps" around $y=x$ as shown over $0 \le x \le 2\pi$.

UNIT 1 REVIEW

1. $\tilde{x} = \frac{\pi}{3}$. Therefore,

$$\cos\frac{5\pi}{3} = \cos\frac{\pi}{3} = \frac{1}{2}$$

2. $\tilde{x} = \frac{\pi}{6}$. Therefore,

$$\csc\frac{11\pi}{6} = -\csc\frac{\pi}{6} = -2$$

3. $\tilde{x} \approx 1.28$. Therefore,

$$\sin 1.86 \approx \sin 1.28 \approx 0.9580$$

4. $\tilde{x} \approx 0.64$. Therefore,

$$\cot 5.64 \approx -\cot 0.64 = -1.343$$

5. $\tilde{x} \approx 0.45$. Therefore,

$$\tan(-5.83) \approx \tan 0.45 = 0.4831$$

6. $\cos(-8.54) \approx \cos(-8.54 + 6.28)$
$= \cos(-2.26)$
For $x = -2.26$, $\tilde{x} \approx 0.88$. Hence

$$\cos(-8.54) \approx -\cos 0.88 \approx -0.6372$$

7. $\sec 10.38 \approx \sec(10.38 - 6.28)$
$= \sec 4.10$
For $x = 4.10$, $\tilde{x} \approx 0.96$. Hence

$$\sec 10.38 \approx -\sec 0.96 \approx -1.744$$

8. $\sin 15.48 \approx \sin(15.48 - 2 \cdot 6.28)$
$= \sin 2.92$
For $x = 2.92$, $\tilde{x} \approx 0.22$. Hence

$$\sin 15.48 \approx \sin 0.22 \approx 0.2182$$

9. $d = -6\cos(4)(2) = -6\cos 8$
$\approx -6\cos(8 - 6.28) = -6\cos 1.72$
$\approx -6(-\cos 1.42) \approx -6(-0.1502)$
≈ 0.90

10. $d = 20\sin 2\pi(1.4) \approx 20\sin 8.79$
$\approx 20\sin(8.79 - 6.28) \approx 20\sin 2.51$
$\approx 20\sin 0.63 \approx 20(0.5891)$
≈ 11.78

11. The period $p = \frac{2\pi}{|\frac{1}{2}|} = 4\pi$. We graph a sine wave and "stretch" each cycle of it over an interval 4π units long. The graph of $y = \sin x$ (dashed curve) is drawn for reference.

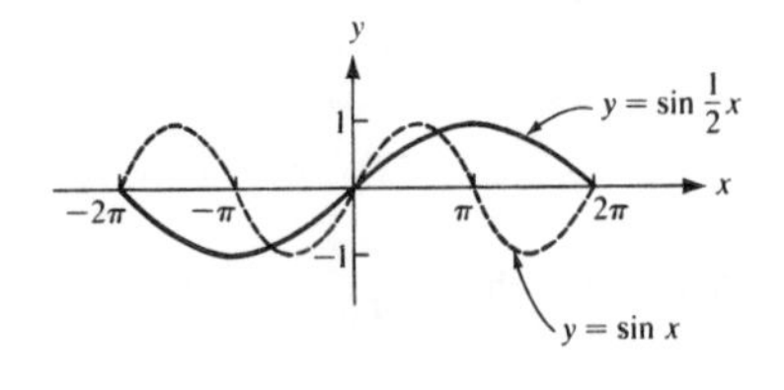

12. The phase shift is π to the right. We graph a cosine wave shifted π units to the right of the graph of $y = \cos x$ (dashed curve), which is drawn for reference.

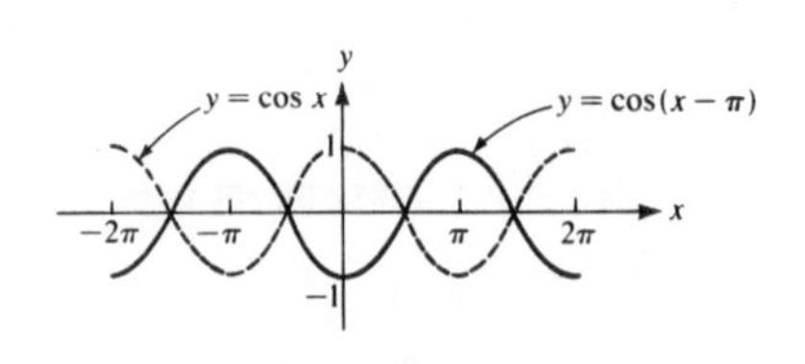

13. From Property 3 of Section 1.2, we note the following things about this sine wave:

1. It has amplitude 2.

2. It has period $\frac{2\pi}{|2|} = \pi$.

3. It is $\frac{\pi}{3}$ units to the left of the graph of $y = 2\sin 2x$.

First we sketch the graph of $y = 2\sin 2x$ (dashed curve), and then with the above facts, we can sketch the curve shown.

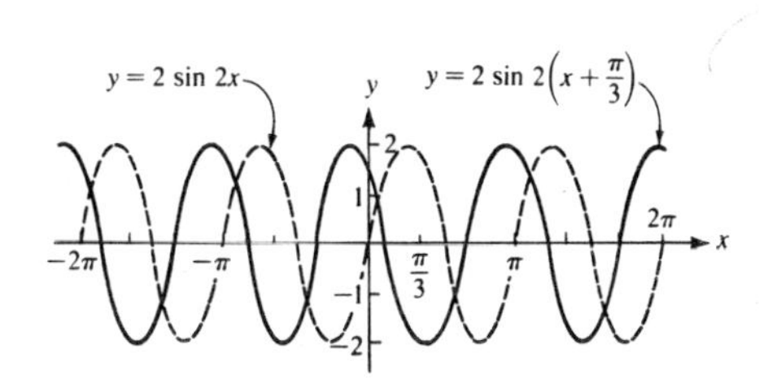

14. From Property 3 of Section 1.2, we note the following things about this cosine wave:

1. It has amplitude $\frac{1}{2}$.

2. It has period $\frac{2\pi}{|\pi/4|} = 8$.

Since the period is 8, we use integers as elements of the domain. Scaling the x axis in integral units facilitates sketching the graph.

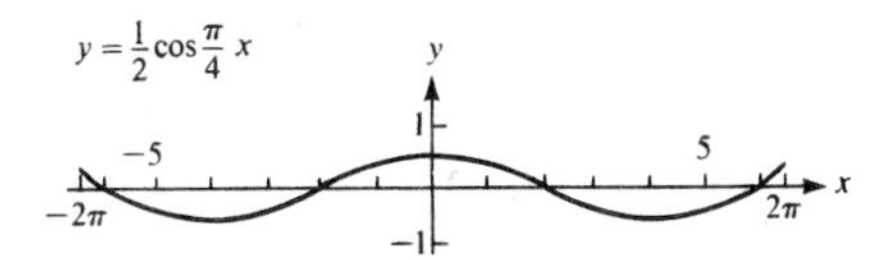

15. See the graph of $y = 2 \sin 2\left(x + \frac{\pi}{3}\right)$ above. The zeros are the x intercepts

$$\left\{-\frac{11\pi}{6}, -\frac{4\pi}{3}, -\frac{5\pi}{6}, -\frac{\pi}{3}, \frac{\pi}{6}, \frac{2\pi}{3}, \frac{7\pi}{6}, \frac{5\pi}{3}\right\}$$

16. See the graph of $y = \frac{1}{2} \cos \frac{\pi}{4} x$ above. The zeros are the x intercepts $\{-6, -2, 2, 6\}$.

17. Since $\csc \frac{2}{3} x$ is undefined for $\frac{2}{3} x = k \cdot \pi$, $k \in J$, there are vertical asymptotes at $x = k \cdot \frac{3\pi}{2}$, $k \in J$. Taking $k = -1, 0, 1$, we obtain vertical asymptotes at $x = -\frac{3\pi}{2}, 0, \frac{3\pi}{2}$.

The period p is given by $p = \frac{2}{\frac{2}{3}} = 3\pi$ and each cycle is "spread out" over an interval of 3π units.

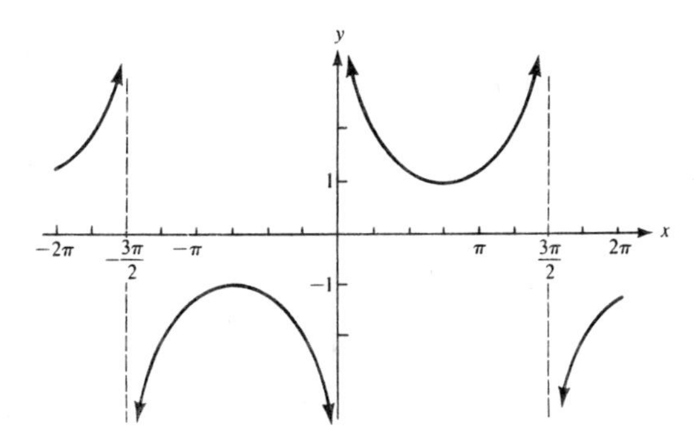

18. Since sec $2x$ is undefined for $2x=\frac{\pi}{2}+k\cdot\pi$, $k\in J$, there are vertical asymptotes at $x=\frac{\pi}{4}+k\cdot\frac{\pi}{2}$, $k\in J$. Thus there are vertical asymptotes at $x=-\frac{7\pi}{4},-\frac{5\pi}{4},-\frac{3\pi}{4},-\frac{\pi}{4},\frac{\pi}{4},\frac{3\pi}{4},\frac{5\pi}{4},\frac{7\pi}{4}$. The period is given by $p=\frac{2\pi}{|2|}=\pi$ and each cycle is "contracted" into an interval of π units. The coefficient $\frac{1}{2}$ gives the low and high points.

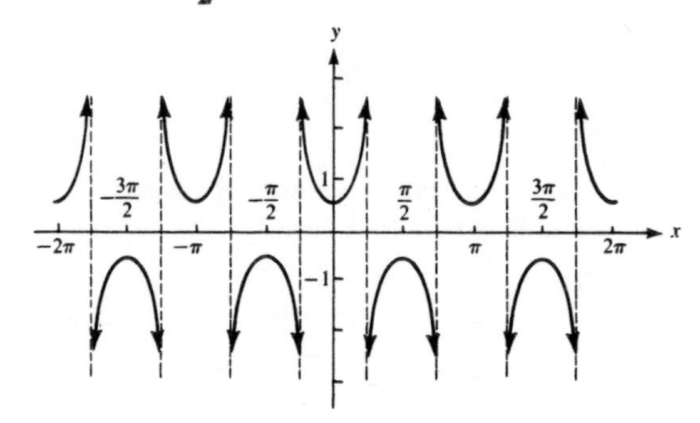

19. First, sketch the graph of $y=\tan x$. Then, shift it $\frac{\pi}{3}$ units to the right to obtain the graph of $y=\tan\left(x-\frac{\pi}{3}\right)$.

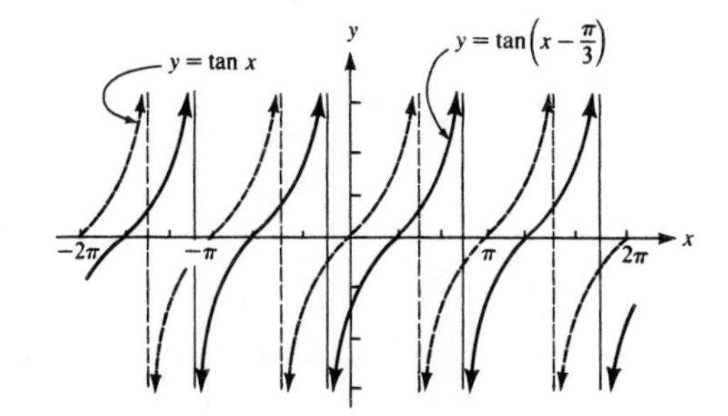

20. First, sketch the graph of $y=\cot x$. Then, shift it $\frac{\pi}{4}$ units to the left to obtain the graph of $y=\cot\left(x+\frac{\pi}{4}\right)$.

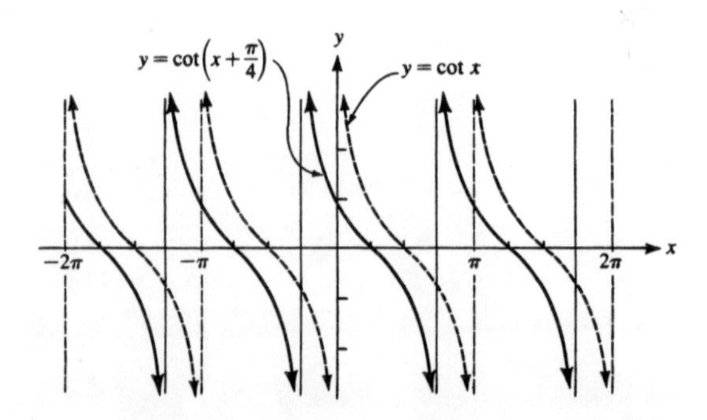

21. First, sketch the graphs of $y = 2 \sin x$ and $y = -\cos 2x$ on the same axes. Then use the method of addition of ordinates.

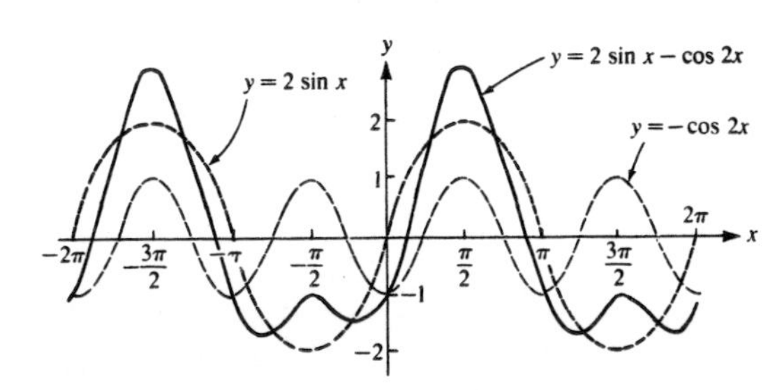

22. First, sketch the graphs of $y = x$ and $y = \sin x$ on the same axes. Then use the method of addition of ordinates.

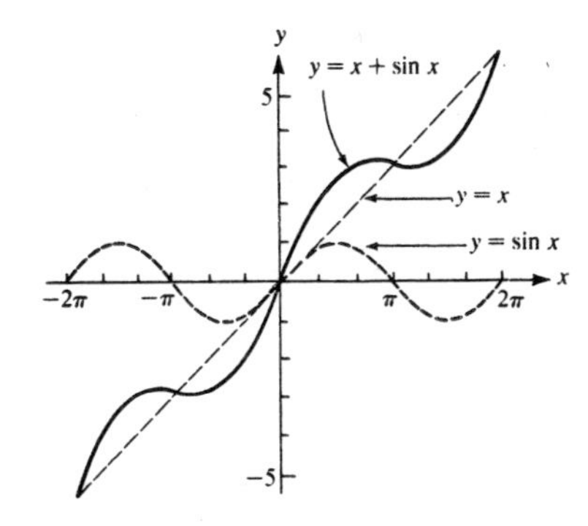

[2.1]

Note: Each of the following identities can be verified in more than one way. Due to limitations of space, only one verification will be shown.

1.

$$\cos x \tan x = \sin x$$
$$\cos x \left(\frac{\sin x}{\cos x}\right) = \sin x$$
$$\sin x = \sin x$$

3.

$$\sin x \cot x = \cos x$$
$$\sin x \left(\frac{\cos x}{\sin x}\right) = \cos x$$
$$\cos x = \cos x$$

5.

$$\sin^2 x \cot^2 x = \cos^2 x$$
$$\sin^2 x \left(\frac{\cos^2 x}{\sin^2 x}\right) = \cos^2 x$$
$$\cos^2 x = \cos^2 x$$

7.

$$\cos^2 x\,(1 + \tan^2 x) = 1$$
$$\cos^2 x + \cos^2 x \tan^2 x = 1$$
$$\cos^2 x + \cos^2 x \left(\frac{\sin^2 x}{\cos^2 x}\right) = 1$$
$$\cos^2 x + \sin^2 x = 1$$
$$1 = 1$$

9.

$$\sin x \sec x = \tan x$$

$$\sin x \left(\frac{1}{\cos x}\right) = \frac{\sin x}{\cos x}$$

$$\frac{\sin x}{\cos x} = \frac{\sin x}{\cos x}$$

11.

$$\sec x - \cos x = \sin x \tan x$$

$$\frac{1}{\cos x} - \cos x = \sin x \left(\frac{\sin x}{\cos x}\right)$$

$$\frac{1}{\cos x} - \frac{\cos^2 x}{\cos x} = \frac{\sin^2 x}{\cos x}$$

$$\frac{1 - \cos^2 x}{\cos x} = \frac{\sin^2 x}{\cos x}$$

$$\frac{\sin^2 x}{\cos x} = \frac{\sin^2 x}{\cos x}$$

13.

$$\sec^2 x - 1 = \frac{1}{\csc^2 x - 1}$$

$$(\tan^2 x + 1) - 1 = \frac{1}{(\cot^2 x + 1) - 1}$$

$$\tan^2 x = \frac{1}{\cot^2 x}$$

$$\tan^2 x = \tan^2 x$$

15.

$$\sec x \csc x - \cot x = \tan x$$

$$\left(\frac{1}{\cos x}\right)\left(\frac{1}{\sin x}\right) - \frac{\cos x}{\sin x} = \tan x$$

$$\frac{1}{\sin x \cos x} - \frac{\cos x}{\sin x} = \tan x$$

$$\frac{1 - \cos^2 x}{\sin x \cos x} = \tan x$$

$$\frac{\sin^2 x}{\sin x \cos x} = \tan x$$

$$\frac{\sin x}{\cos x} = \tan x$$

$$\tan x = \tan x$$

17.

$$\frac{1 + \tan^2 x}{\tan^2 x} = \csc^2 x$$

$$\frac{1}{\tan^2 x} + \frac{\tan^2 x}{\tan^2 x} = \csc^2 x$$

$$\cot^2 x + 1 = \csc^2 x$$

$$\csc^2 x = \csc^2 x$$

19.

$$\tan^2 x - \sin^2 x = \sin^2 x \tan^2 x$$

$$\frac{\sin^2 x}{\cos^2 x} - \sin^2 x = \sin^2 x \frac{\sin^2 x}{\cos^2 x}$$

$$\frac{\sin^2 x}{\cos^2 x} - \frac{\sin^2 x \cos^2 x}{\cos^2 x} = \sin^2 x \frac{\sin^2 x}{\cos^2 x}$$

$$\frac{\sin^2 x - \sin^2 x \cos^2 x}{\cos^2 x} = \frac{\sin^4 x}{\cos^2 x}$$

?

$$\frac{\sin^2 x (1 - \cos^2 x)}{\cos^2 x} = \frac{\sin^4 x}{\cos^2 x}$$

$$\frac{\sin^2 x \sin^2 x}{\cos^2 x} = \frac{\sin^4 x}{\cos^2 x}$$

$$\frac{\sin^4 x}{\cos^2 x} = \frac{\sin^4 x}{\cos^2 x}$$

21. $\tan x + \sec x = \dfrac{1}{\sec x - \tan x}$

$\tan x + \sec x$

$= \dfrac{1}{\sec x - \tan x} \cdot \dfrac{\sec x + \tan x}{\sec x + \tan x}$

$\tan x + \sec x = \dfrac{\sec x + \tan x}{\sec^2 x - \tan^2 x}$

From Property 2, $\sec^2 x - \tan^2 x = 1$. Therefore,

$\tan x + \sec x = \dfrac{\sec x + \tan x}{1}$

$\tan x + \sec x = \tan x + \sec x$

23. $\dfrac{1 - \sec x}{\cos x - 1} = \sec x$

$\dfrac{1 - \dfrac{1}{\cos x}}{\cos x - 1} = \dfrac{1}{\cos x}$

$\dfrac{\cos x \left(1 - \dfrac{1}{\cos x}\right)}{\cos x (\cos x - 1)} = \dfrac{1}{\cos x}$

$\dfrac{\cos x - 1}{\cos x (\cos x - 1)} = \dfrac{1}{\cos x}$

$\dfrac{1}{\cos x} = \dfrac{1}{\cos x}$

25. $\dfrac{\cos x}{\cot x - 6 \cos x} = \dfrac{1}{\csc x - 6}$

$\dfrac{\cos x}{\dfrac{\cos x}{\sin x} - 6 \cos x} = \dfrac{1}{\dfrac{1}{\sin x} - 6}$

$\dfrac{\cos x}{\dfrac{\cos x - 6 \cos x \sin x}{\sin x}} = \dfrac{1}{\dfrac{1 - 6 \sin x}{\sin x}}$

$\dfrac{\sin x \cos x}{\cos x - 6 \cos x \sin x} = \dfrac{\sin x}{1 - 6 \sin x}$

$\dfrac{\cos x \sin x}{\cos x (1 - 6 \sin x)} = \dfrac{\sin x}{1 - 6 \sin x}$

$\dfrac{\sin x}{1 - 6 \sin x} = \dfrac{\sin x}{1 - 6 \sin x}$

27. $\cos x = \dfrac{2 \sin x + 3}{3 \sec x + 2 \tan x}$

$\cos x = \dfrac{2 \sin x + 3}{\dfrac{3}{\cos x} + \dfrac{2 \sin x}{\cos x}}$

$\cos x = \dfrac{(2 \sin x + 3) \cos x}{\left(\dfrac{3}{\cos x} + \dfrac{2 \sin x}{\cos x}\right) \cos x}$

$\cos x = \dfrac{(3 + 2 \sin x) \cos x}{3 + 2 \sin x}$

$\cos x = \cos x$

29.

$$\begin{aligned}
\tan^4 x - \sec^4 x &= 1 - 2 \sec^2 x \\
(\tan^2 x + \sec^2 x)(\tan^2 x - \sec^2 x) &= 1 - 2 \sec^2 x \\
(\sec^2 x - 1 + \sec^2 x)(\sec^2 x - 1 - \sec^2 x) &= 1 - 2 \sec^2 x \\
(2 \sec^2 x - 1)(-1) &= 1 - 2 \sec^2 x \\
1 - 2 \sec^2 x &= 1 - 2 \sec^2 x
\end{aligned}$$

31. From the figure, $\sin x = PB$ and $\cos x = OB$. Hence $\sin^2 x = (PB)^2$ and $\cos^2 x = (OB)^2$. Then, $\sin^2 x + \cos^2 x = (PB)^2 + (OB)^2$. Apply the Pythagorean theorem to right triangle OBP and obtain $(OB)^2 + (PB)^2 = 1$. Thus $\sin^2 x + \cos^2 x = 1$.

33. For $x \neq k$, $k \in J$, divide both members of $\cos^2 x + \sin^2 x = 1$ by $\sin^2 x$ and obtain $\dfrac{\cos^2 x}{\sin^2 x} + \dfrac{\sin^2 x}{\sin^2 x} = \dfrac{1}{\sin^2 x}$, from which $\cot^2 x + 1 = \csc^2 x$.

[2.2]

1. If $\cos x = -\frac{3}{5}$, then depending on whether x is in Quadrant II or III, we have from Property 1 on page 22,

$$\sin x = \pm\sqrt{1-\cos^2 x} = \pm\sqrt{1-\left(-\frac{3}{5}\right)^2} = \pm\sqrt{\frac{16}{25}} = \pm\frac{4}{5}$$

(*a*) If $\frac{\pi}{2} \leq x \leq \pi$, then $\sin x = \frac{4}{5}$, and from Property 5 on page 26,

$$\cos 2x = \cos^2 x - \sin^2 x = \left(-\frac{3}{5}\right)^2 - \left(\frac{4}{5}\right)^2 = \frac{9}{25} - \frac{16}{25} = -\frac{7}{25}$$

(*b*) If $\pi \leq x \leq \frac{3\pi}{2}$, then $\sin x = -\frac{4}{5}$, and from Property 6,

$$\sin 2x = 2\sin x\cos x = 2\left(-\frac{4}{5}\right)\left(-\frac{3}{5}\right) = \frac{24}{25}$$

(*c*) If $3\pi \leq x \leq \frac{7\pi}{2}$, then $\frac{3\pi}{2} \leq \frac{x}{2} \leq \frac{7\pi}{4}$, so that $\frac{x}{2}$ terminates in Quadrant IV, and we have from Property 8,

$$\sin\frac{1}{2}x = -\sqrt{\frac{1-\cos x}{2}} = -\sqrt{\frac{1+\frac{3}{5}}{2}} = -\frac{2\sqrt{5}}{5}$$

(*d*) If $\frac{\pi}{2} \leq x \leq \pi$, then $\frac{\pi}{4} \leq \frac{x}{2} \leq \frac{\pi}{2}$, so that $\frac{x}{2}$ terminates in Quadrant I, and we have from Property 7,

$$\cos\frac{1}{2}x = \sqrt{\frac{1+\cos x}{2}} = \sqrt{\frac{1+\left(-\frac{3}{5}\right)}{2}} = \sqrt{\frac{1-\frac{3}{5}}{2}} = \frac{\sqrt{5}}{5}$$

3. If $\cos x = 0.7$, then depending on the quadrant of x,

$$\sin x = \pm\sqrt{1-\cos^2 x} = \pm\sqrt{1-0.7^2} = \pm\sqrt{1-0.49} = \pm\sqrt{0.51} \approx \pm 0.71$$

(*a*) If $0 \leq x \leq \frac{\pi}{2}$, then $\sin x$ is positive, and

$$\sin 2x = 2\sin x\cos x \approx 2(0.71)(0.7) \approx 0.99$$

(*b*) If $\frac{3\pi}{2} \leq x \leq 2\pi$, $\sin x$ is negative, and

$$\cos 2x = \cos^2 x - \sin^2 x \approx (0.7)^2 - (-0.71)^2 \approx -0.01$$

(*c*) If $0 \le x \le \frac{\pi}{2}$, then $0 \le \frac{x}{2} \le \frac{\pi}{4}$,

so that $\frac{x}{2}$ terminates in Quadrant I,

$$\sin \frac{1}{2}x = \sqrt{\frac{1-\cos x}{2}} = \sqrt{\frac{1-0.7}{2}}$$

$$= \sqrt{0.15} \approx 0.39$$

(*d*) If $\frac{3\pi}{2} \le x \le 2\pi$, then $\frac{3\pi}{4} \le \frac{x}{2} \le \pi$,

so that $\frac{x}{2}$ terminates in Quadrant II,

$$\cos \frac{1}{2}x = -\sqrt{\frac{1+\cos x}{2}}$$

$$= -\sqrt{\frac{1+0.7}{2}}$$

$$= -\sqrt{0.85} \approx -0.92$$

5.

$$\sin 2x = \frac{2\tan x}{1+\tan^2 x}$$

$$2\sin x\cos x = \frac{\frac{2\sin x}{\cos x}\cdot\cos^2 x}{\left(1+\frac{\sin^2 x}{\cos^2 x}\right)\cdot\cos^2 x}$$

$$2\sin x\cos x = \frac{2\sin x\cos x}{\cos^2 x+\sin^2 x}$$

$$2\sin x\cos x = 2\sin x\cos x$$

7.

$$\cot x - \cot 2x = \csc 2x$$

$$\frac{1}{\tan x} - \frac{1}{\tan 2x} = \frac{1}{\sin 2x}$$

$$\frac{1}{\tan x} - \frac{1}{\frac{2\tan x}{1-\tan^2 x}} = \frac{1}{2\sin x\cos x}$$

$$\frac{2}{2\tan x} - \frac{1-\tan^2 x}{2\tan x} = \frac{1}{2\sin x\cos x}$$

$$\frac{2-(1-\tan^2 x)}{2\tan x} = \frac{1}{2\sin x\cos x}$$

$$\frac{1+\tan^2 x}{2\tan x} = \frac{1}{2\sin x\cos x}$$

$$\frac{\sec^2 x}{2\tan x} = \frac{1}{2\sin x\cos x}$$

$$\frac{\frac{1}{\cos^2 x}}{\frac{2\sin x}{\cos x}} = \frac{1}{2\sin x\cos x}$$

$$\frac{1}{\cos^2 x}\cdot\frac{\cos x}{2\sin x} = \frac{1}{2\sin x\cos x}$$

$$\frac{1}{2\sin x\cos x} = \frac{1}{2\sin x\cos x}$$

9.

$$\frac{1+\cos 2x}{\sin 2x} = \cot x$$

$$\frac{1+1-2\sin^2 x}{2\sin x\cos x} = \cot x$$

$$\frac{2-2\sin^2 x}{2\sin x\cos x} = \cot x$$

$$\frac{2(1-\sin^2 x)}{2\sin x\cos x} = \cot x$$

$$\frac{\cos^2 x}{\sin x\cos x} = \cot x$$

$$\frac{\cos x}{\sin x} = \cot x$$

$$\cot x = \cot x$$

11.

$$\begin{aligned}\sin(\pi+x) &= \sin\pi\cos x+\cos\pi\sin x\\ &= 0\cos x+(-1)\sin x\\ &= -\sin x\end{aligned}$$

13.

$$\begin{aligned}\cos(\pi-x) &= \cos\pi\cos x+\sin\pi\sin x\\ &= -1(\cos x)+0(\sin x)\\ &= -\cos x\end{aligned}$$

15. $\cos(2\pi - x) = \cos 2\pi \cos x + \sin 2\pi \sin x$
$= 1(\cos x) + 0 \cdot \sin x$
$= \cos x$

17. $\tan(\pi + x) = \dfrac{\tan \pi + \tan x}{1 - \tan \pi \tan x}$
$= \dfrac{0 + \tan x}{1 - 0(\tan x)}$
$= \tan x$

19. $\sin(-x) = \sin(0 - x)$
$= \sin 0 \cos x - \cos 0 \sin x$
$= 0(\cos x) - 1(\sin x)$
$= -\sin x$

21. $\cos \dfrac{2\pi}{3} = \cos\left(\pi - \dfrac{\pi}{3}\right)$
$= -\cos \dfrac{\pi}{3} = -\dfrac{1}{2}$

23. $\sin \dfrac{5\pi}{3} = \sin\left(2\pi - \dfrac{\pi}{3}\right)$
$= -\sin \dfrac{\pi}{3} = -\dfrac{\sqrt{3}}{2}$

25. $\tan \dfrac{7\pi}{4} = \tan\left(2\pi - \dfrac{\pi}{4}\right)$
$= -\tan \dfrac{\pi}{4} = -1$

27. $\sin 225° = \sin(180° + 45°)$
$= -\sin 45° = -\dfrac{1}{\sqrt{2}}.$

29. $\tan 120° = \tan(180° - 60°),$
$= -\tan 60° = -\sqrt{3}$

31. $\sin(-45°) = -\sin 45° = -\dfrac{1}{\sqrt{2}}$

33. From the given hint we have, by the distance formula for all $x_1, x_2 \in R$,

$$(P_2P_5)^2 = \left(\sqrt{[\cos(x_1 + x_2) - 1]^2 + [\sin(x_1 + x_2) - 0]^2}\right)^2$$
$$(P_1P_4)^2 = \left(\sqrt{(\cos x_2 - \cos x_1)^2 + [\sin x_2 - (-\sin x_1)]^2}\right)^2$$

Since $(P_2P_5)^2 = (P_1P_4)^2$ we have

$$[\cos(x_1 + x_2) - 1]^2 + [\sin(x_1 + x_2)]^2 = (\cos x_2 - \cos x_1)^2 + (\sin x_2 + \sin x_1)^2$$

$$\cos^2(x_1 + x_2) - 2\cos(x_1 + x_2) + 1 + \sin^2(x_1 + x_2)$$
$$= \cos^2 x_2 - 2\cos x_1 \cos x_2 + \cos^2 x_1 + \sin^2 x_2 + 2\sin x_1 \sin x_2 + \sin^2 x_1$$

Regrouping terms,

$$[\cos^2(x_1 + x_2) + \sin^2(x_1 + x_2)] - 2\cos(x_1 + x_2) + 1$$
$$= (\cos^2 x_2 + \sin^2 x_2) + (\cos^2 x_1 + \sin^2 x_1) - 2\cos x_1 \cos x_2 + 2\sin x_1 \sin x_2$$

Since $\cos^2 x + \sin^2 x = 1$,

$$1 - 2\cos(x_1 + x_2) + 1 = 1 + 1 - 2\cos x_1 \cos x_2 + 2\sin x_1 \sin x_2$$
$$2 - 2\cos(x_1 + x_2) = 2 - 2\cos x_1 \cos x_2 + 2\sin x_1 \sin x_2$$
$$-2\cos(x_1 + x_2) = -2\cos x_1 \cos x_2 + 2\sin x_1 \sin x_2$$
$$\cos(x_1 + x_2) = \cos x_1 \cos x_2 - \sin x_1 \sin x_2$$

35. (*a*) In $\cos(x_1 - x_2) = \cos x_1 \cos x_2 + \sin x_1 \sin x_2$, we replace x_1 by $\frac{\pi}{2}$ and x_2 by x and obtain

$$\cos\left(\frac{\pi}{2} - x\right) = \cos\frac{\pi}{2}\cos x + \sin\frac{\pi}{2}\sin x.$$

$$= 0 \cdot \cos x + 1 \cdot \sin x = \sin x$$

(*b*) From (*a*) above, $\sin x = \cos\left(\frac{\pi}{2} - x\right)$. We replace x by $\frac{\pi}{2} - x$ and obtain

$$\sin\left(\frac{\pi}{2} - x\right) = \cos\left[\frac{\pi}{2} - \left(\frac{\pi}{2} - x\right)\right] = \cos\left[\left(\frac{\pi}{2} - \frac{\pi}{2}\right) + x\right] = \cos x$$

(*c*) In $\sin x = \cos\left(\frac{\pi}{2} - x\right)$ we replace x by $x_1 + x_2$ and obtain

$$\sin(x_1 + x_2) = \cos\left[\frac{\pi}{2} - (x_1 + x_2)\right] = \cos\left[\left(\frac{\pi}{2} - x_1\right) - x_2\right]$$

Expanding $\cos\left[\left(\frac{\pi}{2} - x_1\right) - x_2\right]$ using Property 1, we have

$$\sin(x_1 + x_2) = \cos\left(\frac{\pi}{2} - x_1\right)\cos x_2 + \sin\left(\frac{\pi}{2} - x_1\right)\sin x_2$$

From (*a*) and (*b*) above, $\cos\left(\frac{\pi}{2} - x_1\right) = \sin x_1$ and $\sin\left(\frac{\pi}{2} - x_1\right) = \cos x_1$. Therefore,

$$\sin(x_1 + x_2) = \sin x_1 \cos x_2 + \cos x_1 \sin x_2$$

37. In Property 1, we substitute x for x_1 and x_2 and obtain

$$\cos(x + x) = \cos x \cos x - \sin x \sin x$$

$$\cos 2x = \cos^2 x - \sin^2 x \qquad (1)$$

From $\cos^2 x + \sin^2 x = 1$, we replace $\cos^2 x$ by $1 - \sin^2 x$ and obtain

$$\cos 2x = (1 - \sin^2 x) - \sin^2 x = 1 - 2\sin^2 x.$$

From $\cos^2 x + \sin^2 x = 1$, we replace $\sin^2 x$ by $1 - \cos^2 x$ in (1) and obtain

$$\cos 2x = \cos^2 x - (1 - \cos^2 x) = 2\cos^2 x - 1$$

39. In $\cos 2x = 2\cos^2 x - 1$ we replace x by $\frac{x}{2}$ and obtain

$$\cos 2\left(\frac{x}{2}\right) = 2\cos^2\frac{x}{2} - 1$$

from which $\cos x = 2\cos^2\frac{x}{2} - 1$. Solving this last equation for $\cos\frac{x}{2}$, we then obtain $\cos\frac{x}{2} = \pm\sqrt{\frac{1 + \cos x}{2}}$, where the + or − sign is chosen to conform with the sign of $\cos\frac{x}{2}$.

41. If $x_1 + x_2 \neq \frac{\pi}{2} + k \cdot \pi$, $k \in J$, then $\tan(x_1 + x_2) = \frac{\sin(x_1 + x_2)}{\cos(x_1 + x_2)}$. By Properties 1 and 3,

$$\tan(x_1 + x_2) = \frac{\sin x_1 \cos x_2 + \cos x_1 \sin x_2}{\cos x_1 \cos x_2 - \sin x_1 \sin x_2}$$

Dividing the numerator and denominator of the right-hand member by $\cos x_1 \cos x_2$ for $x_1, x_2 \neq \frac{\pi}{2} + k \cdot \pi$, $k \in J$,

$$\tan(x_1 + x_2) = \frac{\dfrac{\sin x_1 \cos x_2}{\cos x_1 \cos x_2} + \dfrac{\cos x_1 \sin x_2}{\cos x_1 \cos x_2}}{\dfrac{\cos x_1 \cos x_2}{\cos x_1 \cos x_2} - \dfrac{\sin x_1 \sin x_2}{\cos x_1 \cos x_2}}$$

$$= \frac{\dfrac{\sin x_1}{\cos x_1} + \dfrac{\sin x_2}{\cos x_2}}{1 - \dfrac{\sin x_1}{\cos x_1} \cdot \dfrac{\sin x_2}{\cos x_2}} = \frac{\tan x_1 + \tan x_2}{1 - \tan x_1 \tan x_2}$$

43. In Property 9, for $x \neq \frac{\pi}{4} + \frac{k\pi}{2}$, $k \in J$, we replace x_1 and x_2 by x and obtain

$$\tan(x + x) = \frac{\tan x + \tan x}{1 - \tan x \tan x} = \frac{2 \tan x}{1 - \tan^2 x}$$

[2.3]

1. Since $\cos x > 0$, x is in Quadrant I or IV and we have

(*a*) $\left\{x | x = k \cdot 2\pi^R \pm \frac{\pi^R}{3},\ k \in J\right\}$

(*b*) $\{x | x = k \cdot 360° \pm 60°,\ k \in J\}$

3. Since $\tan x > 0$, x is in Quadrant I or III and we have

(*a*) $\left\{x | x = \frac{\pi^R}{3} + k\pi^R,\ k \in J\right\}$.

(*b*) $\{x | x = 60° + k \cdot 180°,\ k \in J\}$

5. $\sec x = \sqrt{2}$. Since $\sec x > 0$, x is in Quadrant I or IV and we have

(*a*) $\left\{x | x = \frac{\pi^R}{4} + k \cdot 2\pi^R,\ k \in J\right\} \cup \left\{x | x = \frac{7\pi^R}{4} + k \cdot 2\pi^R,\ k \in J\right\}$

(*b*) $\{x | x = 45° + k \cdot 360°,\ k \in J\} \cup \{x | x = 315° + k \cdot 360°,\ k \in J\}$

7. $\sin x = \frac{1}{4} = 0.25$. Since $\sin x > 0$, x is in Quadrant I or II. Hence,

(*a*) from Table III we have
$\{x | x \approx 0.25^R + k \cdot 2\pi^R,\ k \in J\}$ or
$\{x | x \approx 2.89^R + k \cdot 2\pi^R,\ k \in J\}$

(*b*) from Table II we have
$\{x | x \approx 14°30' + k \cdot 360°,\ k \in J\}$ or
$\{x | x \approx 165°30 + k \cdot 360°,\ k \in J\}$

9. $\cot x = \frac{1}{3} \approx 0.333$. Since $\cot x > 0$, x is in Quadrant I or III. Hence

(*a*) from Table III we have
$\{x \mid x \approx 1.25^R + k\pi^R, \quad k \in J\}$

(*b*) from Table II we have
$\{x \mid x \approx 71°40' + k \cdot 180°, \quad k \in J\}$

11. $\sec x = \frac{5}{2} = 2.5$. Since $\sec x > 0$, x is in Quadrant I or IV. Hence

(*a*) from Table III we have
$\{x \mid x \approx k \cdot 2\pi^R \pm 1.16^R, \quad k \in J\}$

(*b*) from Table II we have
$\{x \mid x \approx k \cdot 360° \pm 66°30', \quad k \in J\}$

13. First noting that $\sin x \neq 0$ in any solution, we divide each member by $\sin x$ to obtain

$$1 - \sqrt{3}\,\frac{\cos x}{\sin x} = 0$$
$$1 - \sqrt{3} \cot x = 0$$
$$\cot x = \frac{1}{\sqrt{3}}$$

Therefore, the solution set is $\left\{x \mid x = \frac{\pi^R}{3} + k \cdot \pi^R, \quad k \in J\right\}$.

15. $\tan x = \pm\sqrt{1} = \pm 1$. Therefore, the solution set is

$$\left\{x \mid x = \frac{\pi^R}{4} + k \cdot \pi^R, \quad k \in J\right\} \cup \left\{x \mid x = \frac{3\pi^R}{4} + k \cdot \pi^R, \quad k \in J\right\}$$

17. $\sin^2 x - \cos^2 x = 1$. Since $\sin^2 x = 1 - \cos^2 x$,

$$1 - \cos^2 x - \cos^2 x = 1$$
$$-2\cos^2 x = 0$$
$$\cos^2 x = 0$$
$$\cos x = 0$$

Therefore, the solution set is $\left\{x = \frac{\pi^R}{2} + k \cdot \pi^R, \quad k \in J\right\}$.

19. If $(2 \sin \theta - 1)(2 \sin^2 \theta - 1) = 0$, then either $2 \sin \theta - 1 = 0$ or $2 \sin^2 \theta - 1 = 0$. Hence either $\sin \theta = \frac{1}{2}$ or $\sin \theta = \pm \frac{1}{\sqrt{2}}$. Therefore the solution set is

$$\{30°, 150°, 45°, 135°, 225°, 315°\}$$

21. If $2 \sin \theta \cos \theta + \sin \theta = 0$, then $\sin \theta\,(2 \cos \theta + 1) = 0$. Hence either $\sin \theta = 0$ or $2 \cos \theta + 1 = 0$, so that either $\sin \theta = 0$ or $\cos \theta = -\frac{1}{2}$. Hence the solution set is

$$\{0°, 120°, 180°, 240°\}$$

23. If $\tan^2 \alpha - 2 \tan \alpha + 1 = 0$, then $(\tan \alpha - 1)^2 = 0$. Hence $\tan \alpha = 1$. Therefore the solution set is $\left\{\frac{\pi^R}{4}, \frac{5\pi^R}{4}\right\}$.

25. If $\cos^2 \alpha + \cos \alpha = 2$, then $\cos^2 \alpha + \cos \alpha - 2 = 0$ so that $(\cos \alpha - 1)(\cos \alpha + 2) = 0$. But $\cos \alpha + 2 = 0$ for no α. Hence those values of α for which $\cos \alpha - 1 = 0$, alone satisfy the equation. Therefore, the solution set is $\{0^R\}$.

27. If $\sec^2 \alpha + 3 \tan \alpha - 11 = 0$, then $\tan^2 \alpha + 1 + 3 \tan \alpha - 11 = 0$. Hence,

$$\tan^2 \alpha + 3 \tan \alpha - 10 = 0, \quad \text{and} \quad (\tan \alpha + 5)(\tan \alpha - 2) = 0$$

Either $\tan \alpha = -5$ or $\tan \alpha = 2$. From Table III, we have

$$\tan \alpha \approx 2 \quad \text{for} \quad \alpha \approx 1.11^R \quad \text{or} \quad 1.11^R + 3.14^R = 4.25^R$$

and

$$\tan \alpha \approx -5 \quad \text{for} \quad \alpha \approx 1.77^R \quad \text{or} \quad 1.77^R + 3.14^R = 4.91^R$$

Hence, the solution set is $\{1.11^R, 1.77^R, 4.25^R, 4.91^R\}$.

29. From the quadratic formula we have

$$\sin \alpha = \frac{-1 \pm \sqrt{1^2 - (4)(1)(-1)}}{2(1)} = \frac{-1 \pm \sqrt{5}}{2} \approx \frac{-1 \pm 2.236}{2}$$

Hence, $\sin \alpha = -1.618$ or $\sin \alpha = 0.618$. Since, for every α, $|\sin \alpha| \leq 1$, the solution set is $\{0.67^R, 2.47^R\}$.

31. If $\sin 2x - \cos x = 0$, then $2 \sin x \cos x - \cos x = 0$ so that $\cos x(2 \sin x - 1) = 0$. Hence either $\cos x = 0$, in which case $x = \frac{\pi^R}{2} + k \cdot \pi^R$, $k \in J$, or else $2 \sin x - 1 = 0$, in which case $\sin x = \frac{1}{2}$; so that $x = \frac{\pi^R}{6} + k \cdot 2\pi^R$ or $\frac{5\pi^R}{6} + k \cdot 2\pi^R$, $k \in J$. Hence, the solution set is

$$\left\{x \middle| x = \frac{\pi^R}{2} + k \cdot \pi^R,\ k \in J\right\} \cup \left\{x \middle| x = \frac{\pi^R}{6} + k \cdot 2\pi^R,\ k \in J\right\}$$

$$\cup \left\{x \middle| x = \frac{5\pi^R}{6} + k \cdot 2\pi^R,\ k \in J\right\}$$

33. If $\cos 2x = \cos x - 1$, then $2 \cos^2 x - 1 = \cos x - 1$ so that $2 \cos^2 x - \cos x = 0$. Hence, $\cos x(2 \cos x - 1) = 0$. Therefore either $\cos x = 0$ or $2 \cos x - 1 = 0$. Hence either $x = \frac{\pi^R}{2} + k \cdot \pi^R$, $k \in J$, or $\cos x = \frac{1}{2}$ so that

$$x = \frac{\pi^R}{3} + k \cdot 2\pi^R,\ k \in J \quad \text{or} \quad -\frac{\pi^R}{3} + k \cdot 2\pi^R,\ k \in J$$

and the the solution set is

$$\left\{x \middle| x = \frac{\pi^R}{2} + k \cdot \pi^R,\ k \in J\right\} \cup \left\{x \middle| x = \frac{\pi^R}{3} + k \cdot 2\pi^R \text{ or } -\frac{\pi^R}{3} + k \cdot 2\pi^R,\ k \in J\right\}$$

35. If $\cos 2x = \frac{\sqrt{2}}{2}$, then $2x = \frac{\pi}{4} + k \cdot 2\pi$ from which $x = \frac{\pi}{8} + k \cdot \pi$, $k \in J$, or $2x = \frac{7\pi}{4} + k \cdot 2\pi$ from which $x = \frac{7\pi}{8} + k \cdot \pi$, $k \in J$. Thus the solution set is

$$\left\{x \mid x = \frac{\pi}{8} + k \cdot \pi\right\} \cup \left\{x \mid x = \frac{7\pi}{8} + k \cdot \pi\right\}, \quad k \in J$$

37. If $\cos 2x \sin x + \sin x = 0$, then $\sin x(\cos 2x + 1) = 0$. Hence, either $\sin x = 0$ from which $x = k \cdot \pi$, $k \in J$, or $\cos 2x = -1$ from which $2x = \pi + k \cdot 2\pi$, which in turn leads to $x = \frac{\pi}{2} + k \cdot \pi$. Thus the solution set is

$$\{x \mid x = k \cdot \pi, \quad k \in J\} \cup \left\{x \mid x = \frac{\pi}{2} + k \cdot \pi, \quad k \in J\right\}$$

39. If $\sin 4x - 2\sin 2x = 0$, then $2\sin 2x \cos 2x - 2\sin 2x = 0$ or $2\sin 2x(\cos 2x - 1) = 0$. Hence, either $\sin 2x = 0$ from which $2x = 0 + k \cdot \pi$ from which $x = k \cdot \frac{\pi}{2}$, $k \in J$, or $\cos 2x = 1$ and $2x = 0 + k \cdot 2\pi$ from which $x = k \cdot \pi$. Thus the solution set is

$$\left\{x \mid x = k \cdot \frac{\pi}{2}, \quad k \in J\right\} \cup \{x \mid x = k \cdot \pi, \quad k \in J\} = \left\{x \mid x = k \cdot \frac{\pi}{2}, \quad k \in J\right\}$$

[2.4]

1. Let $y = \text{Arcsin}\, \frac{1}{2}$. Then, $\sin y = \frac{1}{2}$, $-\frac{\pi}{2} \leq y \leq \frac{\pi}{2}$. Hence, $y = \frac{\pi}{6}$, and we have $\text{Arcsin}\, \frac{1}{2} = \frac{\pi}{6}$.

3. Let $y = \text{Tan}^{-1}(-\sqrt{3})$. Then $\tan y = -\sqrt{3}$, $-\frac{\pi}{2} < y < \frac{\pi}{2}$. Hence, $y = -\frac{\pi}{3}$ and we have $\text{Tan}^{-1}(-\sqrt{3}) = -\frac{\pi}{3}$.

5. Let $y = \text{Cos}^{-1}\, 2$. Then $\cos y = 2$, $0 \leq y \leq \pi$. Since $|\cos y| \leq 1$ for every y, then $\text{Cos}^{-1}\, 2$ does not exist.

7. Let $y = \text{Arctan}\left(-\frac{1}{\sqrt{3}}\right)$. Then $\tan y = -\frac{1}{\sqrt{3}}$, $-\frac{\pi}{2} < y < \frac{\pi}{2}$. Hence, $y = -\frac{\pi}{6}$ and we have $\text{Arctan}\left(-\frac{1}{\sqrt{3}}\right) = -\frac{\pi}{6}$.

9. Let $y = \text{Arctan } 0.1003$. Then $\tan y = 0.1003$, $-\frac{\pi}{2} < y < \frac{\pi}{2}$. From Table III we find $y \approx 0.10$ and we have $\text{Arctan } 0.1003 \approx 0.10$.

11. Let $y = \text{Sin}^{-1}\ 0.3802$. Then $\sin y = 0.3802$, $-\frac{\pi}{2} \le y \le \frac{\pi}{2}$. From Table III, $y \approx 0.39$ and we have $\text{Sin}^{-1}\ 0.3802 \approx 0.39$.

13. Let $y = \text{Arcsin}\,(-0.8624)$. Then $\sin y = -0.8624$, $-\frac{\pi}{2} \le y \le \frac{\pi}{2}$. From Table III we find that for $\sin y = 0.8624$, $y \approx 1.04$. Hence, for $\sin y = -0.8624$, $-\frac{\pi}{2} \le y \le \frac{\pi}{2}$, we take $y \approx -1.04$ and we have $\text{Arcsin}\,(-0.8624) \approx -1.04$.

15. Let $y = \text{Sin}^{-1}\left(\cos\frac{\pi}{4}\right)$. Then since $\cos\frac{\pi}{4} = \frac{1}{\sqrt{2}}$, $y = \text{Sin}^{-1}\left(\frac{1}{\sqrt{2}}\right)$, $-\frac{\pi}{2} \le y \le \frac{\pi}{2}$. Hence, $\sin y = \frac{1}{\sqrt{2}}$ and $y = \frac{\pi}{4}$. Therefore, $\text{Sin}^{-1}\left(\cos\frac{\pi}{4}\right) = \frac{\pi}{4}$.

17. Let $y = \text{Tan}^{-1}\left(\tan\frac{\pi}{3}\right)$. Then since $\tan\frac{\pi}{3} = \sqrt{3}$, $y = \text{Tan}^{-1}\,(\sqrt{3})$, $-\frac{\pi}{2} < y < \frac{\pi}{2}$. Hence, $\tan y = \sqrt{3}$ and $y = \frac{\pi}{3}$. Therefore, $\text{Tan}^{-1}\left(\tan\frac{\pi}{3}\right) = \frac{\pi}{3}$.

19. Let $\text{Arccos}\,\frac{1}{2} = y$. Then $\cos y = \frac{1}{2}$, $0 \le y \le \pi$. Hence, $y = \frac{\pi}{3}$. Then, $\sin\left(\text{Arccos}\,\frac{1}{2}\right) = \sin\frac{\pi}{3} = \frac{\sqrt{3}}{2}$.

21. Let $\text{Arctan}\,(-1) = y$. Then $\tan y = -1$, $-\frac{\pi}{2} < y < \frac{\pi}{2}$. Hence, $y = \text{Arctan}\,(-1) = -\frac{\pi}{4}$. Then, $\sin\,(\text{Arctan}\,(-1)) = \sin\left(-\frac{\pi}{4}\right) = -\frac{1}{\sqrt{2}}$. Finally,

$$\text{Arccos}\,(\sin\,(\text{Arctan}\,(-1))) = \text{Arccos}\left(-\frac{1}{\sqrt{2}}\right) = \frac{3\pi}{4}$$

23. $3\sin\theta = 2$

$\sin\theta = \frac{2}{3}$

$\theta = \text{Arcsin}\,\frac{2}{3}$, $-\frac{\pi}{2} \le \theta \le \frac{\pi}{2}$

25. $\cos 3\theta = \frac{1}{4}$

$3\theta = \text{Cos}^{-1}\,\frac{1}{4}$, $0 \le 3\theta \le \pi$

$\theta = \frac{1}{3}\,\text{Cos}^{-1}\,\frac{1}{4}$, $0 \le \theta \le \frac{\pi}{3}$

27. $3 \tan 2\theta = 4$

$$\tan 2\theta = \frac{4}{3}$$

$$2\theta = \text{Tan}^{-1}\frac{4}{3}, \quad -\frac{\pi}{2} < 2\theta < \frac{\pi}{2}$$

$$\theta = \frac{1}{2}\text{Tan}^{-1}\frac{4}{3}, \quad -\frac{\pi}{4} < \theta < \frac{\pi}{4}$$

29. $\frac{1}{2}\sin\frac{\theta}{3} = 0.28$

$$\sin\frac{\theta}{3} = 0.56$$

$$\frac{\theta}{3} = \text{Sin}^{-1}\,0.56, \quad -\frac{\pi}{2} \leq \frac{\theta}{3} \leq \frac{\pi}{2}$$

$$\theta = 3\,\text{Sin}^{-1}\,0.56, \quad -\frac{3\pi}{2} \leq \theta \leq \frac{3\pi}{2}$$

31. Let $\alpha = \text{Arccos}(\cos x)$, $0 \leq \alpha \leq \pi$. Then, by definition of Arccosine, $\cos\alpha = \cos x$ and since it is given that $0 \leq x \leq \pi$, $\alpha = x$. Therefore since $\alpha = \text{Arccos}(\cos x)$, we have $\text{Arccos}(\cos x) = x$.

UNIT 2 REVIEW

1. $\cot x = \csc x \cos x$

$$\frac{\cos x}{\sin x} = \frac{1}{\sin x}\cdot\cos x$$

$$\frac{\cos x}{\sin x} = \frac{\cos x}{\sin x}$$

2. $\frac{1}{1+\tan^2 x} = \cos^2 x$

$$\frac{1}{\sec^2 x} = \cos^2 x$$

$$\cos^2 x = \cos^2 x$$

3. $\tan x + \cot x = \csc x \sec x$

$$\frac{\sin x}{\cos x} + \frac{\cos x}{\sin x} = \frac{1}{\sin x}\cdot\frac{1}{\cos x}$$

$$\frac{\sin^2 x + \cos^2 x}{\sin x \cos x} = \frac{1}{\sin x \cos x}$$

$$\frac{1}{\sin x \cos x} = \frac{1}{\sin x \cos x}$$

4. $\frac{1-\sin x}{\cos x} = \frac{\cos x}{1+\sin x}$

$$\left(\frac{1+\sin x}{1+\sin x}\right)\left(\frac{1-\sin x}{\cos x}\right) = \frac{\cos x}{1+\sin x}$$

$$\frac{1-\sin^2 x}{\cos x(1+\sin x)} = \frac{\cos x}{1+\sin x}$$

$$\frac{\cos^2 x}{\cos x(1+\sin x)} = \frac{\cos x}{1+\sin x}$$

$$\frac{\cos x}{1+\sin x} = \frac{\cos x}{1+\sin x}$$

5. $\sec x + \tan x = \frac{1}{\sec x - \tan x}$

$$\sec x + \tan x = \frac{1}{\sec x - \tan x}\cdot\frac{\sec x + \tan x}{\sec x + \tan x}$$

$$\sec x + \tan x = \frac{\sec x + \tan x}{\sec^2 x - \tan^2 x}$$

$$\sec x + \tan x = \frac{\sec x + \tan x}{\tan^2 x + 1 - \tan^2 x}$$

$$\sec x + \tan x = \frac{\sec x + \tan x}{1}$$

$$\sec x + \tan x = \sec x + \tan x$$

6. $2\cos^2 x - 1 = \cos^4 x - \sin^4 x$

$2\cos^2 x - 1$

$\quad = (\cos^2 x - \sin^2 x)(\cos^2 x + \sin^2 x)$

$2\cos^2 x - 1 = (\cos^2 x - \sin^2 x)(1)$

$2\cos^2 x - 1 = \cos^2 x - (1 - \cos^2 x)$

$2\cos^2 x - 1 = 2\cos^2 x - 1$

7. $\tan 2x = \dfrac{2\sin x\cos x}{\cos^2 x - \sin^2 x}$

$\tan 2x = \dfrac{\sin 2x}{\cos 2x}$

$\tan 2x = \tan 2x$

8. $\dfrac{\tan 2x}{2\tan x} = \dfrac{\cot^2 x}{\cot^2 x - 1}$

$$\frac{\dfrac{2\tan x}{1-\tan^2 x}}{2\tan x} = \frac{\cot^2 x}{\cot^2 x - 1}$$

$$\frac{1}{1-\tan^2 x} = \frac{\cot^2 x}{\cot^2 x - 1}$$

$$\frac{1}{1-\dfrac{1}{\cot^2 x}} = \frac{\cot^2 x}{\cot^2 x - 1}$$

$$\frac{1}{\dfrac{\cot^2 x - 1}{\cot^2 x}} = \frac{\cot^2 x}{\cot^2 x - 1}$$

$$\frac{\cot^2 x}{\cot^2 x - 1} = \frac{\cot^2 x}{\cot^2 x - 1}$$

9. If $\cos x = -\dfrac{2}{3}$, then $\sin x = \pm\sqrt{1-\cos^2 x} = \pm\sqrt{1-\left(-\dfrac{2}{3}\right)^2} = \pm\dfrac{\sqrt{5}}{3}$.

(*a*) If $\dfrac{\pi}{2} \le x \le \pi$, then $\sin x = \dfrac{\sqrt{5}}{3}$, and

$$\cos 2x = \cos^2 x - \sin^2 x = \left(-\frac{2}{3}\right)^2 - \left(\frac{\sqrt{5}}{3}\right)^2 = \frac{4}{9} - \frac{5}{9} = -\frac{1}{9}$$

(*b*) If $\pi \le x \le \dfrac{3\pi}{2}$, then $\sin x = -\dfrac{\sqrt{5}}{3}$ and

$$\sin 2x = 2\sin x\cos x = 2\left(-\frac{\sqrt{5}}{3}\right)\left(-\frac{2}{3}\right) = \frac{4\sqrt{5}}{9}$$

(*c*) If $3\pi \le x \le \dfrac{7\pi}{2}$, then $\dfrac{3\pi}{2} \le \dfrac{x}{2} \le \dfrac{7\pi}{4}$, so that $\dfrac{x}{2}$ terminates in Quadrant IV, and we have

$$\sin\frac{1}{2}x = -\sqrt{\frac{1-\cos x}{2}} = -\sqrt{\frac{1+\dfrac{2}{3}}{2}} = -\sqrt{\frac{5}{6}} = -\frac{\sqrt{30}}{6}$$

(*d*) If $\dfrac{\pi}{2} \le x \le \pi$, then $\dfrac{\pi}{4} \le \dfrac{x}{2} \le \dfrac{\pi}{2}$, so that $\dfrac{x}{2}$ terminates in Quadrant I, and we have

$$\cos\frac{1}{2}x = \sqrt{\frac{1+\cos x}{2}} = \sqrt{\frac{1-\dfrac{2}{3}}{2}} = \sqrt{\frac{1}{6}} = \frac{\sqrt{6}}{6}$$

10. If $\sin x = 0.6$, then depending on the quadrant of x,

$$\cos x = \pm\sqrt{1-\sin^2 x} = \pm\sqrt{1-(0.6)^2} = \pm\sqrt{0.64} = \pm 0.8$$

(*a*) If $0 \le x \le \frac{\pi}{2}$, then $\cos x$ is positive and

$$\sin 2x = 2 \sin x \cos x = 2(0.6)(0.8) = 0.96$$

(*b*) If $\frac{3\pi}{2} \le x \le 2\pi$, $\cos x$ is positive and

$$\cos 2x = \cos^2 x - \sin^2 x = (0.8)^2 - (0.6)^2 = 0.28$$

(*c*) If $0 \le x \le \frac{\pi}{2}$, then $\frac{1}{2}x$ terminates in Quadrant I, and

$$\sin \frac{1}{2}x = \sqrt{\frac{1-\cos x}{2}} = \sqrt{\frac{1-0.8}{2}} = \sqrt{\frac{1}{10}} = \frac{\sqrt{10}}{10} \approx 0.32$$

(*d*) If $\frac{3\pi}{2} \le x \le 2\pi$, then $\frac{x}{2}$ terminates in Quadrant II, and

$$\cos \frac{1}{2}x = -\sqrt{\frac{1+\cos x}{2}} = -\sqrt{\frac{1+0.8}{2}} = -\sqrt{0.9} \approx -0.95$$

11. $\sin \frac{5\pi}{4} = \sin\left(\pi + \frac{\pi}{4}\right)$. Using

$\sin(\pi + x) = -\sin x$,

$$\sin \frac{5\pi}{4} = -\sin \frac{\pi}{4} = -\frac{1}{\sqrt{2}}$$

12. $\tan \frac{7\pi}{6} = \tan\left(\pi + \frac{\pi}{6}\right)$. Using

$\tan(\pi + x) = \tan x$,

$$\tan \frac{7\pi}{6} = \tan \frac{\pi}{6} = \frac{1}{\sqrt{3}}$$

13. $\cos \frac{11\pi}{6} = \cos\left(2\pi - \frac{\pi}{6}\right)$. Using

$\cos(2\pi - x) = \cos x$,

$$\cos \frac{11\pi}{6} = \cos \frac{\pi}{6} = \frac{\sqrt{3}}{2}$$

14. $\tan 330° = \tan(360° - 30°)$. Using $\tan(360° - x) = -\tan x$,

$$\tan 330° = -\tan 30° = -\frac{1}{\sqrt{3}}$$

15. Using $\sin(-x) = -\sin x$,

$$\sin(-60°) = -\sin 60° = -\frac{\sqrt{3}}{2}$$

16. Using $\cos(-x) = \cos x$,

$$\cos(-45°) = \cos 45° = \frac{1}{\sqrt{2}}$$

17. (*a*) $\left\{x \mid x = \frac{\pi^R}{6} + k \cdot 2\pi^R,\ k \in J\right\} \cup \left\{x \mid x = \frac{5\pi^R}{6} + k \cdot 2\pi^R,\ k \in J\right\}$

(*b*) $\{x \mid x = 30° + k \cdot 360°,\ k \in J\} \cup \{x \mid x = 150° + k \cdot 360°,\ k \in J\}$

18. (a) $\left\{x \mid x = \frac{5\pi^R}{6} + k \cdot \pi^R, \quad k \in J\right\}$

(b) $\{x \mid x = 150° + k \cdot 180°, \quad k \in J\}$

19. (a) $\{x \mid x \approx 1.32^R + k \cdot 2\pi^R, \quad k \in J\} \cup \{x \mid x \approx 4.96^R + k \cdot 2\pi^R, \quad k \in J\}$

(b) $\{x \mid x \approx 75°30' + k \cdot 360°, \quad k \in J\} \cup \{x \mid x \approx 284°30' + k \cdot 360°, \quad k \in J\}$

20. $3 \sec x + 5 = 0$

$$3 \sec x = -5$$

$$\sec x = -\frac{5}{3} \approx -1.667$$

(a) $\{x \mid x \approx 2.22^R + k \cdot 2\pi, \quad k \in J\} \cup \{x \mid x \approx 4.06^R + k \cdot 2\pi^R, \quad k \in J\}$

(b) $\{x \mid x \approx 126°50' + k \cdot 360°, \quad k \in J\} \cup \{x \mid x \approx 233°10' + k \cdot 360°, \quad k \in J\}$

21. $\sqrt{3} \sin x + \cos x = 0$

$$\sqrt{3}\frac{\sin x}{\cos x} + 1 = 0$$

$$\sqrt{3} \tan x + 1 = 0$$

$$\tan x = -\frac{1}{\sqrt{3}}$$

$\left\{x \mid x = \frac{5\pi^R}{6} + k \cdot \pi^R, \quad k \in J\right\}$

22.

$$\cos^2 x - \sin^2 x = -1$$

$$1 - \sin^2 x - \sin^2 x = -1$$

$$-2 \sin x^2 = -2$$

$$\sin x^2 = 1$$

$$\sin x = \pm 1$$

$\left\{x \mid x = \frac{\pi^R}{2} + k \cdot \pi^R, \quad k \in J\right\}$

23. $\tan \theta = -1$ or $\sin \theta = \frac{1}{2}$

$\{30°, 135°, 150°, 315°\}$

24. $\cos \theta(\tan \theta - 1) = 0$, $\cos \theta = 0$, or $\tan \theta - 1 = 0$

$\{45°, 90°, 225°, 270°\}$

25. $(2 \sin \alpha + 1)(\sin \alpha - 1) = 0$

$\left\{\frac{\pi^R}{2}, \frac{7\pi^R}{6}, \frac{11\pi^R}{6}\right\}$

26.

$$2 \tan^2 \alpha - \tan \alpha - 3 = 0$$

$$(2 \tan \alpha - 3)(\tan \alpha + 1) = 0$$

$\left\{0.98^R, 4.12^R, \frac{3\pi^R}{4}, \frac{7\pi^R}{4}\right\}$

27. Let $y = \text{Arccos}\,\frac{\sqrt{3}}{2}$. Then

$$\cos y = \frac{\sqrt{3}}{2}, \quad 0 \leq y \leq \pi$$

Hence, $\text{Arccos}\,\frac{\sqrt{3}}{2} = \frac{\pi}{6}$.

28. Let $y = \text{Sin}^{-1}\left(-\frac{\sqrt{3}}{2}\right)$. Then

$$\sin y = -\frac{\sqrt{3}}{2}, \quad -\frac{\pi}{2} \leq y \leq \frac{\pi}{2}$$

Hence, $\text{Sin}^{-1}\left(-\frac{\sqrt{3}}{2}\right) = -\frac{\pi}{3}$.

29. Let $y = \text{Arccos } 0.7648$. Then

$$\cos y = 0.7648, \quad 0 \le y \le \pi$$

From Table III, we find that $y \approx 0.70$. Since, $0 \le 0.70 \le \pi$,

$$\text{Arccos } 0.7648 \approx 0.70$$

30. Let $y = \text{Tan}^{-1}(-0.2027)$. Then

$$\tan y = -0.2027, \quad -\frac{\pi}{2} < y < \frac{\pi}{2}$$

Using Table III, we find that

$y \approx -0.20$. Since $-\frac{\pi}{2} < -0.20 < \frac{\pi}{2}$,

$$\text{Tan}^{-1}(-0.2027) \approx -0.20$$

31. Let $y = \text{Cos}^{-1}\left(\sin\frac{\pi}{4}\right)$. $\sin\frac{\pi}{4} = \frac{1}{\sqrt{2}}$.

Therefore, $y = \text{Cos}^{-1}\left(\frac{1}{\sqrt{2}}\right)$ and

$$\cos y = \frac{1}{\sqrt{2}}, \quad 0 \le y \le \pi$$

$y = \frac{\pi}{4}$ and $\text{Cos}^{-1}\left(\sin\frac{\pi}{4}\right) = \frac{\pi}{4}$.

32. Let $y = \text{Tan}^{-1}\left(\tan\frac{\pi}{6}\right)$. $\tan\frac{\pi}{6} = \frac{1}{\sqrt{3}}$.

Therefore $y = \text{Tan}^{-1}\left(\frac{1}{\sqrt{3}}\right)$ and

$$\tan y = \frac{1}{\sqrt{3}}, \quad -\frac{\pi}{2} < y < \frac{\pi}{2}$$

$y = \frac{\pi}{6}$ and $\text{Tan}^{-1}\left(\tan\frac{\pi}{6}\right) = \frac{\pi}{6}$.

33. Let $y = \text{Arccos}\left(-\frac{1}{2}\right)$. Then

$$\cos y = -\frac{1}{2}, \quad 0 \le y \le \pi$$

Hence, $y = \frac{2\pi}{3}$ and

$$\sin\left(\text{Arccos}\left(-\frac{1}{2}\right)\right) = \sin y$$

$$= \sin\frac{2\pi}{3} = \frac{\sqrt{3}}{2}$$

34. Let $y = \text{Sin}^{-1} 0$. Then

$$\sin y = 0, \quad -\frac{\pi}{2} \le y \le \frac{\pi}{2}$$

Hence, $y = 0$ and

$$\cos(\text{Sin}^{-1} 0) = \cos y = \cos 0 = 1$$

35. $\tan\theta = \frac{2}{3}$. Hence

$$\theta = \text{Arctan}\frac{2}{3}, \quad -\frac{\pi}{2} < \theta < \frac{\pi}{2}$$

36. $\cos 3\theta = \frac{3}{4}$

$$3\theta = \text{Cos}^{-1}\frac{3}{4}, \quad 0 \le 3\theta \le \pi$$

$$\theta = \frac{1}{3}\text{Cos}^{-1}\frac{3}{4}, \quad 0 \le \theta \le \frac{\pi}{3}$$

[3.1]

1. $|3 + 2i| = \sqrt{3^2 + 2^2} = \sqrt{13}$

3. $|4 - i| = \sqrt{4^2 + (-1)^2} = \sqrt{17}$

5. $|4| = |4 + 0i| = \sqrt{4^2 + 0^2} = 4$

7. $|3i| = |0 + 3i| = \sqrt{0^2 + 3^2} = 3$

9. $r = |3+3i| = \sqrt{3^2+3^2} = 3\sqrt{2}$. Since $\tan\theta = \frac{3}{3} = 1$ and the graph of $3+3i$ is in Quadrant I, $\theta = 45°$. Hence,

$$3+3i = 3\sqrt{2}(\cos 45° + i\sin 45°) = 3\sqrt{2}\text{ cis }45°$$

11. $r = |5-0i| = \sqrt{5^2} = 5$. Since the graph of $z=5$ is on the positive x axis, $\theta = 0°$. Hence,

$$5 = 5\text{ cis }0°$$

13. $r = |2\sqrt{3}-2i| = \sqrt{(2\sqrt{3})^2+(-2)^2} = 4$. Since $\tan\theta = \frac{-2}{2\sqrt{3}} = -\frac{1}{\sqrt{3}}$ and the graph of $2\sqrt{3}-2i$ is in Quadrant IV, $\theta = 330°$. Hence, $2\sqrt{3}-2i = 4\text{ cis }330°$.

15. Since $\cos 240° = -\frac{1}{2}$ and $\sin 240° = -\frac{\sqrt{3}}{2}$, we have from Property 4

$$4\text{ cis }240° = 4(\cos 240° + i\sin 240°) = 4\left(-\frac{1}{2}-\frac{\sqrt{3}}{2}i\right) = -2-2\sqrt{3}i$$

17. Since $\cos(-30°) = \frac{\sqrt{3}}{2}$ and $\sin(-30°) = -\frac{1}{2}$, we have from Property 4

$$6\text{ cis }(-30°) = 6[\cos(-30°) + i\sin(-30°)] = 6\left(\frac{\sqrt{3}}{2}-\frac{1}{2}i\right) = 3\sqrt{3}-3i$$

19. Since $\cos 420° = \frac{1}{2}$ and $\sin 420° = \frac{\sqrt{3}}{2}$, we have

$$12\text{ cis }420° = 12(\cos 420° + i\sin 420°) = 12\left(\frac{1}{2}+\frac{\sqrt{3}}{2}i\right) = 6+6\sqrt{3}\,i$$

21. (*a*) By Property 5,

$$z_1\cdot z_2 = (3\text{ cis }90°)(\sqrt{2}\text{ cis }45°) = 3\cdot\sqrt{2}\text{ cis }(90°+45°) = 3\sqrt{2}\text{ cis }135°$$

Since $\cos 135° = -\frac{1}{\sqrt{2}}$ and $\sin 135° = \frac{1}{\sqrt{2}}$,

$$3\sqrt{2}\text{ cis }135° = 3\sqrt{2}\left(-\frac{1}{\sqrt{2}}+\frac{1}{\sqrt{2}}i\right) = -3+3i$$

(*b*) By Property 6,

$$\frac{z_1}{z_2} = \frac{3\text{ cis }90°}{\sqrt{2}\text{ cis }45°} = \frac{3}{\sqrt{2}}\text{ cis }(90°-45°) = \frac{3}{\sqrt{2}}\text{ cis }45°$$

Since $\cos 45° = \frac{1}{\sqrt{2}}$ and $\sin 45° = \frac{1}{\sqrt{2}}$,

$$\frac{3}{\sqrt{2}}\text{ cis }45° = \frac{3}{\sqrt{2}}\left(\frac{1}{\sqrt{2}}+\frac{1}{\sqrt{2}}i\right) = \frac{3}{2}+\frac{3}{2}i$$

23. (*a*) By Property 5,

$$z_1 \cdot z_2 = (6 \text{ cis } 150°)(18 \text{ cis } 570°) = 6 \cdot 18 \text{ cis } (150° + 570°) = 108 \text{ cis } 720°.$$

Since $\cos 720° = 1$ and $\sin 720° = 0$,

$$108 \text{ cis } 720° = 108(1 + 0i) = 108$$

(*b*) By Property 6,

$$\frac{z_1}{z_2} = \frac{6 \text{ cis } 150°}{18 \text{ cis } 570°} = \frac{1}{3} \text{ cis } (150° - 570°) = \frac{1}{3} \text{ cis } (-420°)$$

Since $\cos(-420°) = \frac{1}{2}$ and $\sin(-420°) = -\frac{\sqrt{3}}{2}$,

$$\frac{z_1}{z_2} = \frac{1}{3} \text{ cis } (-420°) = \frac{1}{3}\left(\frac{1}{2} - \frac{\sqrt{3}}{2} i\right) = \frac{1}{6} - \frac{\sqrt{3}}{6} i$$

25. $z_1 = -3 + i \approx \sqrt{10}(\text{cis } 161°30')$
$z_2 = -2 - 4i \approx 2\sqrt{5}(\text{cis } 243°30')$

(*a*) $z_1 z_2 \approx (\sqrt{10} \cdot 2\sqrt{5})(\text{cis } (161°30' + 243°30'))$

$$= 2\sqrt{50} \text{ cis } 405° = 10\sqrt{2}\left(\frac{1}{\sqrt{2}} + \frac{1}{\sqrt{2}} i\right) = 10 + 10i$$

(*b*) $$\frac{z_1}{z_2} = \frac{\sqrt{10} \text{ cis } 161°30'}{2\sqrt{5} \text{ cis } 243°30'} = \frac{\sqrt{2}}{2} \text{ cis } (-82°)$$

$$= \frac{\sqrt{2}}{2} \text{ cis } 82° \approx \frac{\sqrt{2}}{2}(0.139 + 0.990i)$$

$$\approx 0.707(0.139 + 0.990i) \approx 0.10 + 0.70i$$

27. $-\frac{1}{2} - \frac{i\sqrt{3}}{2} = 1 \cdot \text{cis } 240°$. Then, from an extension of Property 5:

$$[1 \cdot \text{cis } (240°)]^3 = [1 \cdot 1 \cdot 1 \cdot \text{cis } (240° + 240° + 240°)] = \text{cis } 720° = 1$$

Hence, $\left(-\frac{1}{2} - \frac{i\sqrt{3}}{2}\right)^3 = \text{cis } 720° = 1.$

29. Let $z = a + bi$. Then $\bar{z} = a - bi$. We then have

$$z + \bar{z} = (a + bi) + (a - bi) = (a + a) + (b - b)i = 2a$$

Since $2a$ is a real number, we have the desired result.

[3.2]

1. By De Moivre's Theorem we have

$$[2 \text{ cis } (-30°)]^7 = 2^7 \text{ cis } [(7)(-30°)] = 128 \text{ cis } (-210°)$$

Since $\cos(-210°) = -\frac{\sqrt{3}}{2}$ and $\sin(-210°) = \frac{1}{2}$, then

$$128 \text{ cis } (-210°) = 128\left(-\frac{\sqrt{3}}{2} + \frac{1}{2}i\right) = -64\sqrt{3} + 64i$$

3. $\left(-\frac{1}{2} + \frac{1}{2}\sqrt{3}\,i\right)^3 = (1 \text{ cis } 120°)^3 = 1^3 \text{ cis } 360° = 1$

5. $(\sqrt{3} \text{ cis } 5°)^{12} = 3^6 \text{ cis } 60° = 3^6\left(\frac{1}{2} + \frac{\sqrt{3}}{2}i\right) = \frac{729}{2} + \frac{729\sqrt{3}}{2}i$

7. $(\sqrt{3} - i)^{-5} = \left[2\left(\frac{\sqrt{3}}{2} - \frac{1}{2}i\right)\right]^{-5} = [2 \text{ cis } 330°]^{-5}$

$$= 2^{-5} \text{ cis } (-1650°) = 2^{-5} \text{ cis } (-1650° + 5 \cdot 360°)$$

$$= 2^{-5} \text{ cis } 150° = \frac{1}{32}\left(-\frac{\sqrt{3}}{2} + \frac{1}{2}i\right) = -\frac{\sqrt{3}}{64} + \frac{1}{64}i$$

9. $(\text{cis } 60°)^{-3} = (1 \text{ cis } 60°)^{-3} = (1^{-3}) \text{ cis } [(-3)(60°)] = 1 \cdot \text{cis } (-180°)$. Since

$\cos(-180°) = -1$ and $\sin(-180°) = 0$, then $\text{cis } (-180°) = (-1 + 0i) = -1$

11. By Property 2, each number

$$w_k = 32^{1/5} \text{ cis } \left(\frac{45° + k \cdot 360°}{5}\right), \quad k \in J$$

is a fifth root of z. Taking $32^{1/5} = 2$ and $k = 0, 1, 2, 3$, and 4 in turn gives the roots

$$w_0 = 2 \text{ cis } 9°, \quad w_1 = 2 \text{ cis } 81°, \quad w_2 = 2 \text{ cis } 153°, \quad w_3 = 2 \text{ cis } 225°, \quad w_4 = 2 \text{ cis } 297°$$

13. If $z = -16\sqrt{3} + 16i$, then in trigonometric form, $z = 32 \text{ cis } 150°$. By Property 2, each number

$$w_k = 32^{1/5} \text{ cis } \left(\frac{150° + k \cdot 360°}{5}\right), \quad k \in J$$

is a fifth root of z. Hence

$$w_0 = 2 \text{ cis } 30°, \quad w_1 = 2 \text{ cis } 102°, \quad w_2 = 2 \text{ cis } 174°, \quad w_3 = 2 \text{ cis } 246°, \quad w_4 = 2 \text{ cis } 318°$$

15. If $z = 0 - 1i$, then $z = 1 \text{ cis } 270°$. By Property 2, each number

$$w_k = 1^{1/6} \text{ cis } \left(\frac{270° + k \cdot 360°}{6}\right), \quad k \in J$$

is a sixth root of z. Hence

$$w_0 = 1 \text{ cis } 45°, \quad w_1 = \text{cis } 105°, \quad w_2 = \text{cis } 165°,$$
$$w_3 = \text{cis } 225°, \quad w_4 = \text{cis } 285°, \quad w_5 = \text{cis } 345°$$

17. We seek the five fifth roots of $16 - 16\sqrt{3}\,i = 32 \text{ cis } (-60°)$. These are the numbers

$$w_k = 32^{1/5} \text{ cis } \left(\frac{-60° + k \cdot 360°}{5}\right), \quad k \in J$$

Thus we have

$$w_0 = 2 \text{ cis } (-12°), \quad w_1 = 2 \text{ cis } 60°, \quad w_2 = 2 \text{ cis } 132°, \quad w_3 = 2 \text{ cis } 204°, \quad w_4 = 2 \text{ cis } 276°$$

19. We seek the seven seventh roots of $-1 + 0i = 1 \cdot \text{cis } (180°)$. These are the numbers

$$w_k = 1^{1/7} \text{ cis } \left(\frac{180° + k \cdot 360°}{7}\right), \quad k \in J$$

Thus we have

$$w_0 = \text{cis } \frac{180°}{7}, \quad w_1 = \text{cis } \frac{540°}{7}, \quad w_2 = \text{cis } \frac{900°}{7} \quad w_3 = \text{cis } \frac{1260°}{7},$$

$$w_4 = \text{cis } \frac{1620°}{7}, \; w_5 = \text{cis } \frac{1980°}{7}, \quad w_6 = \text{cis } \frac{2340°}{7}.$$

[3.3]

1. The figures show the possibilities.

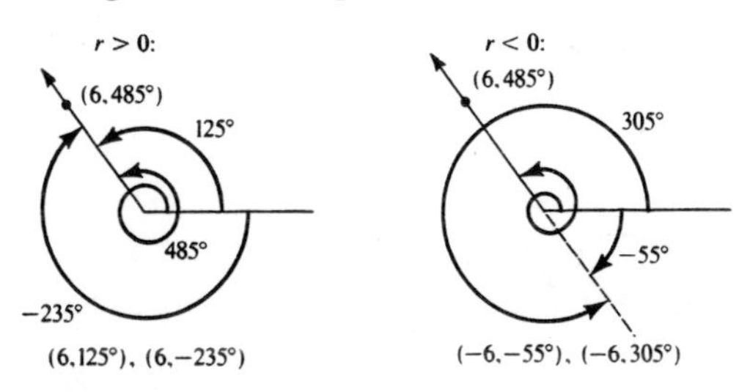

3. (2,90°),(2,−270°), (−2,−90°), (−2,270°)

5. (6,120°), (6,−240°), (−6,−60°), (−6,300°)

7. From Property 2a,

$$x = 5 \cos 45° = 5\left(\frac{1}{\sqrt{2}}\right) = \frac{5}{\sqrt{2}}$$

$$y = 5 \sin 45° = 5\left(\frac{1}{\sqrt{2}}\right) = \frac{5}{\sqrt{2}}$$

Hence, $(x,y) = \left(\frac{5}{\sqrt{2}}, \frac{5}{\sqrt{2}}\right)$.

9. $x=\frac{1}{2}\cos 330^\circ=\frac{1}{2}\left(\frac{\sqrt{3}}{2}\right)=\frac{\sqrt{3}}{4}$

$y=\frac{1}{2}\sin 330^\circ=\frac{1}{2}\left(-\frac{1}{2}\right)=-\frac{1}{4}$

Hence, $(x,y)=\left(\frac{\sqrt{3}}{4},-\frac{1}{4}\right)$.

11. $x=10\cos(-135^\circ)=10\left(-\frac{1}{\sqrt{2}}\right)=-\frac{10}{\sqrt{2}}=-5\sqrt{2}$

$y=10\sin(-135^\circ)=10\left(-\frac{1}{\sqrt{2}}\right)=-5\sqrt{2}$

Hence, $(x,y)=(-5\sqrt{2},-5\sqrt{2})$.

13. From Property 2b, $r=\pm\sqrt{(3\sqrt{2})^2+(3\sqrt{2})^2}=\pm\sqrt{18+18}=\pm 6$. Choosing the positive sign and noting that the point $(3\sqrt{2},3\sqrt{2})$ is in Quadrant I, we have from Property 2b, $\tan\theta=\frac{3\sqrt{2}}{3\sqrt{2}}=1$, from which $\theta=45^\circ$. Hence, a pair of polar coordinates is $(6,45^\circ)$. A pair with an angle of negative measure is $(6,-315^\circ)$.

15. From Property 2b, $r=\pm\sqrt{(-1)^2+(-\sqrt{3})^2}=\pm\sqrt{4}=\pm 2$. Choosing the positive sign and noting that the point $(-1,-\sqrt{3})$ is in Quadrant III, we have $\tan\theta=\frac{-\sqrt{3}}{-1}=\sqrt{3}$, from which $\theta=240^\circ$. Hence, a pair of polar coordinates is $(2,240^\circ)$. A pair with an angle of negative measure is $(2,-120^\circ)$.

17. For the origin, $r=0$ and we can specify any angle θ. We choose $(0,0^\circ)$, $(0,-60^\circ)$.

19. Substituting $x=r\cos\theta$ and $y=r\sin\theta$, we have

$$r^2\cos^2\theta+r^2\sin^2\theta=25$$
$$r^2(\cos^2\theta+\sin^2\theta)=25$$
$$r^2(1)=25$$

Hence, we have $r=5$ or $r=-5$.

21. Substituting $y=r\sin\theta$, we have $r\sin\theta=-4$.

23. Substituting $x=r\cos\theta$ and $y=r\sin\theta$,

$$r^2\cos^2\theta+9r^2\sin^2\theta=9$$
$$r^2(\cos^2\theta+9\sin^2\theta)=9$$

25. Substituting from Property 2b, $\pm\sqrt{x^2+y^2}=5$. Squaring each side, $x^2+y^2=25$.

27. From Property 2a, we substitute $\frac{x}{r}$ for $\cos\theta$ and obtain $r=\frac{9x}{r}$ from which $r^2=9x$. Then, since $r^2=x^2+y^2$, we have $x^2+y^2=9x$.

29. $r(1 - \cos\theta) = 2$

$r - r\cos\theta = 2$

From Property 2a, $r\cos\theta = x$ and from Property 2b, $r = \pm\sqrt{x^2 + y^2}$. Hence,

$$\pm\sqrt{x^2+y^2} - x = 2$$
$$\pm\sqrt{x^2+y^2} = x + 2$$

Squaring both members and simplifying, we obtain $y^2 = 4x + 4$.

31.
$$r = \sec^2\left(\frac{\theta}{2}\right) = \frac{1}{\cos^2\left(\frac{\theta}{2}\right)}; \quad r\cos^2\left(\frac{\theta}{2}\right) = 1$$

Using the fact that $\cos\frac{\theta}{2} = \pm\sqrt{\frac{1+\cos\theta}{2}}$ (Property 7 of Section 2.2), and squaring each member, $\cos^2\frac{\theta}{2} = \frac{1+\cos\theta}{2}$. Substituting the right-hand member in $r\cos^2\left(\frac{\theta}{2}\right) = 1$ yields

$$r\left(\frac{1+\cos\theta}{2}\right) = 1$$
$$r(1+\cos\theta) = 2$$
$$\pm\sqrt{x^2+y^2}\left(1 + \frac{x}{\pm\sqrt{x^2+y^2}}\right) = 2$$
$$\pm\sqrt{x^2+y^2} + x = 2$$
$$\pm\sqrt{x^2+y^2} = 2 - x$$
$$x^2 + y^2 = x^2 - 4x + 4$$
$$x = -\frac{1}{4}y^2 + 1 \quad \text{which is an equation of a parabola.}$$

33. First we make the table below.

θ	$\sin\theta$	r or $9\sin\theta$	(r,θ)
0°	0	0	(0,0°)
30°	0.5	4.5	(4.5,30°)
45°	0.7	6.3	(6.3,45°)
60°	0.9	8.1	(8.1,60°)
90°	1.0	9.0	(9.0,90°)
120°	0.9	8.1	(8.1,120°)
135°	0.7	6.3	(6.3,135°)
150°	0.5	4.5	(4.5,150°)
180°	0	0	(0,180°)

Continuing the table would produce no points not already obtained from the table. Graphing each (r,θ) leads us to the graph on page 112.

35. First we make the table below.

θ	$\cos\theta$	$2\cos\theta$	r or $1+2\cos\theta$	(r,θ)
0°	1.0	2.0	3.0	(3.0,0°)
30°	0.9	1.8	2.8	(2.8,30°)
45°	0.7	1.4	2.4	(2.4,45°)
60°	0.5	1.0	2.0	(2.0,60°)
90°	0.0	0.0	1.0	(1.0,90°)
120°	−0.5	−1.0	0.0	(0.0,120°)
135°	−0.7	−1.4	−0.4	(−0.4,135°)
150°	−0.9	−1.8	−0.8	(−0.8,150°)
180°	−1.0	−2.0	−1.0	(−1.0,180°)

We continue in this manner for $180° < \theta < 360°$ and then graph each (r,θ).

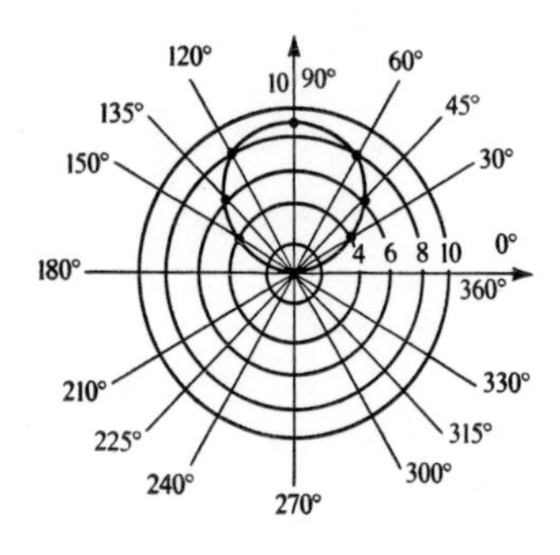

Ex. 33

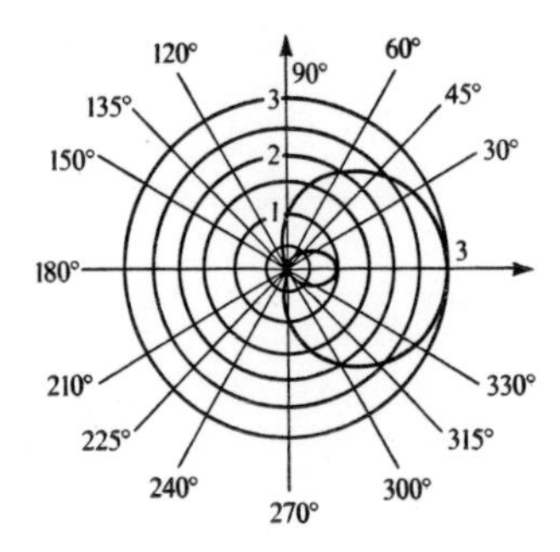

Ex. 35

37. First we make the table below.

θ	$\sin\theta$	r or $1-\sin\theta$	(r,θ)
0°	0	1.0	(1.0,0°)
30°	0.5	0.5	(0.5,30°)
45°	0.7	0.3	(0.3,45°)
60°	0.9	0.1	(0.1,60°)
90°	1.0	0.0	(0.0,90°)
120°	0.9	0.1	(0.1,120°)
135°	0.7	0.3	(0.3,135°)
150°	0.5	0.5	(0.5,150°)
180°	0.0	1.0	(1.0,180°)

We continue in this manner for $180° < \theta < 360°$ and then graph each (r,θ) as shown on page 113.

39. First we make the table below.

θ	3θ	$\sin 3\theta$	$4 \sin 3\theta$	(r,θ)
0°	0°	0.0	0.0	(0,0°)
10°	30°	0.5	2.0	(2.0,10°)
15°	45°	0.7	2.8	(2.8,15°)
20°	60°	0.9	3.5	(3.5,20°)
30°	90°	1.0	4.0	(4.0,30°)
40°	120°	0.9	3.5	(3.5,40°)
50°	150°	0.5	2.0	(2.0,50°)
60°	180°	0	0	(0,60°)

We continue in this manner for $60° < \theta < 360°$.

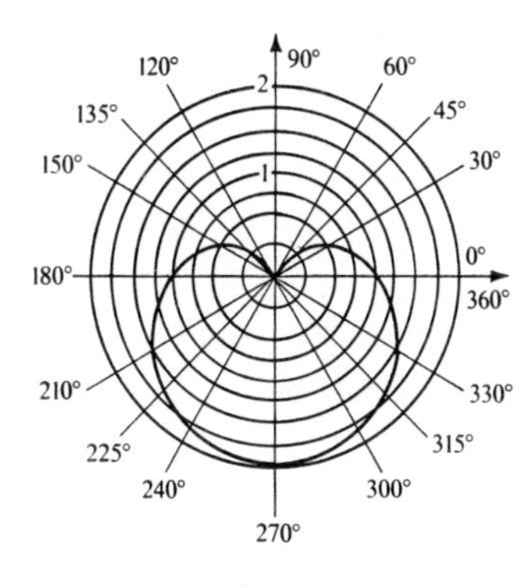

Ex. 37

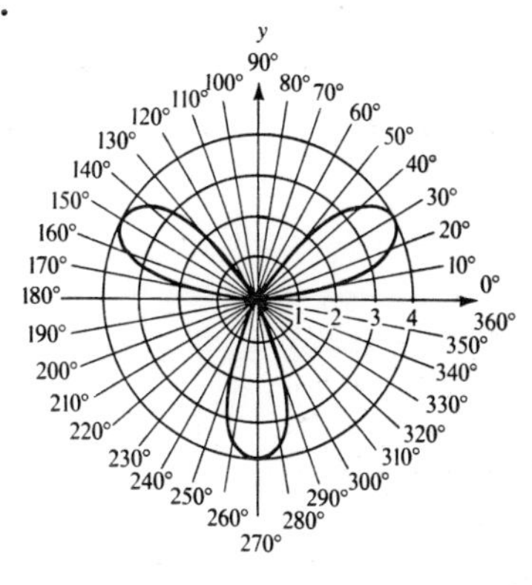

Ex. 39

UNIT 3 REVIEW

1. $|3-4i| = \sqrt{(3)^2+(-4)^2} = 5$

2. $|-5| = \sqrt{(-5)^2+0^2} = 5$

3. $|2+3i| = \sqrt{(2)^2+(3)^2} = \sqrt{13}$

4. $|-3i| = \sqrt{(0)^2+(-3)^2} = 3$

5. $r = |3-3i| = \sqrt{(3)^2+(-3)^2} = 3\sqrt{2}$

Since $\tan\theta = \dfrac{-3}{3} = -1$ and the graph of $3-3i$ is in Quadrant IV, $\theta = 315°$. Hence, $3-3i = 3\sqrt{2}\text{ cis } 315°$.

6. $r = |-2i| = \sqrt{(0)^2+(-2)^2} = 2$

From the graph of $-2i$, we take $\theta = 270°$. Hence $-2i = 2\text{ cis } 270°$.

7. $a = 2\cos 300° = 2\left(\dfrac{1}{2}\right) = 1$

$b = 2\sin 300° = 2\left(\dfrac{-\sqrt{3}}{2}\right) = -\sqrt{3}$

and $2\text{ cis } 300° = 1-\sqrt{3}\,i$.

8. $a = 6\cos(-60°) = 6\left(\dfrac{1}{2}\right) = 3$

$b = 6\sin(-60°) = 6\left(-\dfrac{\sqrt{3}}{2}\right) = -3\sqrt{3}$

Hence $6\text{ cis }(-60°) = 3-3\sqrt{3}i$.

9. (a) $z_1 \cdot z_2 = (4 \text{ cis } 60°)(2 \text{ cis } 60°)$
$= 4 \cdot 2 \text{ cis } (60° + 60°),$
$= 8 \cos 120° + i(8 \sin 120°)$

$$= 8\left(-\frac{1}{2}\right) + 8\left(\frac{\sqrt{3}}{2}\right)i$$

$= -4 + 4\sqrt{3}\,i$

(b) $\dfrac{z_1}{z_2} = \dfrac{4 \text{ cis } 60°}{2 \text{ cis } 60°} = 2 \text{ cis } (60° - 60°)$
$= 2 \cos 0° + i(2 \sin 0°) = 2$

10. $z_1 \approx \sqrt{10} \text{ cis } 341°30',$
$z_2 \approx 2\sqrt{5} \text{ cis } 63°30'$

(a) $z_1 \cdot z_2 \approx \sqrt{10}{\cdot}2\sqrt{5}$
$\text{cis } (341°30' + 63°30')$
$= 2\sqrt{50} \text{ cis } 405° = 10\sqrt{2} \text{ cis } 405°$

$$= 10\sqrt{2}\left(\frac{1}{\sqrt{2}} + \frac{1}{\sqrt{2}}i\right) = 10 + 10i$$

(b) $\dfrac{z_1}{z_2} \approx \dfrac{\sqrt{10} \text{ cis } 341°30'}{2\sqrt{5} \text{ cis } 63°30'}$

$$= \frac{\sqrt{10}}{2\sqrt{5}} \text{ cis } (341°30' - 63°30')$$

$$= \frac{\sqrt{2}}{2} \text{ cis } 278° \approx 0.707(+0.139 - 0.990i)$$

$\approx 0.707(+0.139 - 0.990i) \approx 0.10 - 0.70i$

11. $[2 \text{ cis } (-60°)]^5 = 2^5 \text{ cis } [5 \cdot (-60°)]$
$= 32(\cos (-300°) + i \sin (-300°))$

$$= 32\left(\frac{1}{2} + \frac{\sqrt{3}}{2}i\right) = 16 + 16\sqrt{3}\,i$$

12. $(1 - i)^{10} = (\sqrt{2} \text{ cis}(-45°))^{10}$
$= (\sqrt{2})^{10} \text{ cis } [10 \cdot (-45°)]$
$= 32(\cos (-450°) + i \sin (-450°))$
$= 32(0 - i) = -32i$

13. $(\sqrt{3} + i)^4 = (2 \text{ cis } 30°)^4 = 2^4 \text{ cis } (4 \cdot 30°)$

$$= 16\left(-\frac{1}{2} + \frac{\sqrt{3}}{2}i\right)$$

$= -8 + 8\sqrt{3}\,i$

14. $(-1 + i)^{-6} = (\sqrt{2} \text{ cis } 135°)^6$
$= (\sqrt{2})^6 \text{ cis } (6 \cdot 135°)$
$= 8(\cos 810° + i \sin 810°)$
$= 8(\cos 90° + i \sin 90°)$
$= 8(0 + i) = 8i$

15. $w_k = 16^{1/4} \text{ cis } \left(\dfrac{60° + k \cdot 360°}{4}\right), \quad k \in J$

$w_0 = 2 \text{ cis } 15°, \quad w_1 = 2 \text{ cis } 105°, \quad w_2 = 2 \text{ cis } 195°, \quad w_3 = 2 \text{ cis } 285°$

16. In trigonometric form, $z = 32 \text{ cis } 60°.$

$$w_k = 32^{1/5} \text{ cis } \left(\frac{60° + k \cdot 360°}{5}\right), \quad k \in J$$

$w_0 = 2 \text{ cis } 12°, \quad w_1 = 2 \text{ cis } 84°, \quad w_2 = 2 \text{ cis } 156°, \quad w_3 = 2 \text{ cis } 228°, \quad w_4 = 2 \text{ cis } 300°$

17. We seek the five fifth roots of $16\sqrt{3} + 16i = 32 \text{ cis } 30°$. These are the numbers

$$w_k = 32^{1/5} \text{ cis } \left(\frac{30° + k \cdot 360°}{5}\right), \quad k \in J$$

$w_0 = 2 \text{ cis } 6°, \quad w_1 = 2 \text{ cis } 78°, \quad w_2 = 2 \text{ cis } 150°, \quad w_3 = 2 \text{ cis } 222°, \quad w_4 = 2 \text{ cis } 294°$

18. We seek the six sixth roots of $1 = \text{cis } 0°$. These are the numbers

$$w_k = 1^{1/6} \text{ cis} \left(\frac{0° + k \cdot 360°}{6}\right), \quad k \in J$$

$w_0 = \text{cis } 0°$, $w_1 = \text{cis } 60°$, $w_2 = \text{cis } 120°$, $w_3 = \text{cis } 180°$, $w_4 = \text{cis } 240°$, $w_5 = \text{cis } 300°$

19. Since the angles whose measures are 140° and −220° are coterminal with the angle whose measure is 500°, we have (4,140°), (4,−220°). Taking $r < 0$, we obtain (−4,−40°), (−4,320°).

20. Since the angles whose measures are 20° and −340° are coterminal with the angle whose measure is −700°, we have (−3,20°), (−3,−340°). Taking $r > 0$, we obtain (3,−160°), (3,200°).

21. $x = -2 \cos 60° = -2\left(\frac{1}{2}\right) = -1$

$y = -2 \sin 60° = -2\left(\frac{\sqrt{3}}{2}\right) = -\sqrt{3}$

Hence the rectangular coordinates are $(-1, -\sqrt{3})$.

22. $x = 6 \cos(-225°) = 6\left(-\frac{\sqrt{2}}{2}\right) = -3\sqrt{2}$

$y = 6 \sin(-225°) = 6\left(\frac{\sqrt{2}}{2}\right) = 3\sqrt{2}$

Hence the rectangular coordinates are $(-3\sqrt{2}, 3\sqrt{2})$.

23. $r = \pm\sqrt{(2\sqrt{2})^2 + (2\sqrt{2})^2} = \pm 4$ and we choose the positive sign. Then, noting that the point $(2\sqrt{2}, 2\sqrt{2})$ is in the first quadrant,

$$\tan \theta = \frac{2\sqrt{2}}{2\sqrt{2}} = 1$$

from which $\theta = 45°$. Hence a pair of coordinates is (4,45°). A pair with an angle of negative measure is (4,−315°).

24. $r = \pm\sqrt{(1)^2 + (-\sqrt{3})^2} = \pm 2$, and we choose the positive sign. Then, noting that the point $(1, -\sqrt{3})$ is in Quadrant IV,

$$\tan \theta = \frac{-\sqrt{3}}{1} = -\sqrt{3}$$

from which $\theta = 300°$. Hence, a pair of coordinates is (2,300°). A pair with an angle of negative measure is (2,−60°).

25. Using $x = r \cos \theta$ and $y = r \sin \theta$,

$$9r^2 \cos^2 \theta + r^2 \sin^2 \theta = 9$$

26. Using $x = r \cos \theta$ and $y = r \sin \theta$,

$$9r^2 \cos^2 \theta - r^2 \sin^2 \theta = 9$$

27. Using $r = \pm\sqrt{x^2 + y^2}$, we have $\pm\sqrt{x^2 + y^2} = 2$. Squaring each side, $x^2 + y^2 = 4$.

28. Using $r = \pm\sqrt{x^2 + y^2}$ and $x = r\cos\theta$,

$$\pm\sqrt{x^2 + y^2} + x = 2$$
$$\pm\sqrt{x^2 + y^2} = 2 - x$$

Squaring each member,

$$x^2 + y^2 = 4 - 4x + x^2$$
$$y^2 + 4x - 4 = 0$$

29. First we make the table below.

θ	$\sin\theta$	r or $2\sin\theta$	(r,θ)
0°	0.0	0.0	(0,0°)
30°	0.5	1.0	(1.0,30°)
45°	0.7	1.4	(1.4,45°)
60°	0.9	1.8	(1.8,60°)
90°	1.0	2.0	(2.0,90°)
120°	0.9	1.8	(1.8,120°)
135°	0.7	1.4	(1.4,135°)
150°	0.5	1.0	(1.0,150°)
180°	0.0	0.0	(0,180°)

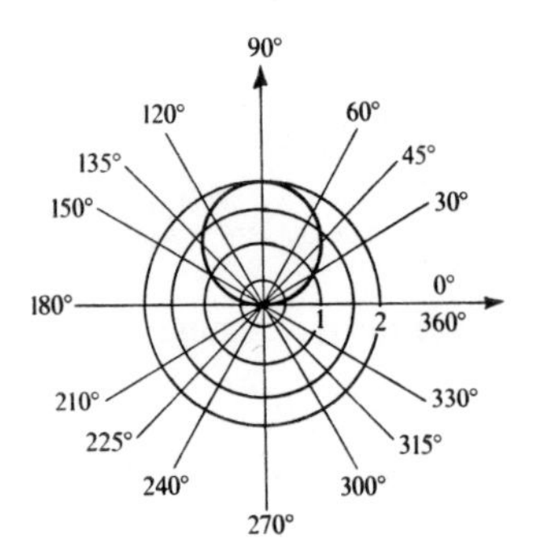

Continuing the table would yield no new points.

30. First we make the table below.

θ	$\sin\theta$	r or $1 + \sin\theta$	(r,θ)
0°	0.0	1.0	(1.0,0°)
30°	0.5	1.5	(1.5,30°)
45°	0.7	1.7	(1.7,45°)
60°	0.9	1.9	(1.9,60°)
90°	1.0	2.0	(2.0,90°)
120°	0.9	1.9	(1.9,120°)
135°	0.7	1.7	(1.7,135°)
150°	0.5	1.5	(1.5,150°)
180°	0.0	1.0	(1.0,180°)
210°	−0.5	0.5	(0.5,210°)
225°	−0.7	0.3	(0.3,225°)
240°	−0.9	0.1	(0.1,240°)
270°	−1.0	0.0	(0.0,270°)
300°	−0.9	0.1	(0.1,300°)
315°	−0.7	0.3	(0.3,315°)
330°	−0.5	0.5	(0.5,330°)
360°	0.0	1.0	(1.0,360°)

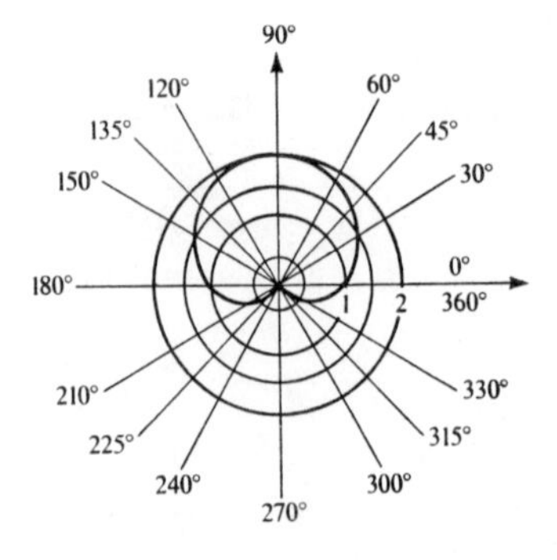

[A.1]

1. $\sqrt{4}\sqrt{-1} = 2i$

3. $\sqrt{16}\sqrt{-1}\sqrt{2} = 4i\sqrt{2}$

5. $3\sqrt{4}\sqrt{-1}\sqrt{2} = 6i\sqrt{2}$

7. $4 + 2i$

9. $3\sqrt{25}\sqrt{-1}\sqrt{2} + 2 = 15i\sqrt{2} + 2$
$= 2 + 15i\sqrt{2}$

11. $2 + \sqrt{4}\sqrt{-1} = 2 + 2i$

13. $\frac{1}{3}\sqrt{-1}\sqrt{25} = \frac{1}{3}i(5) = \frac{5}{3}i$

15. $\frac{\sqrt{36}\sqrt{2}}{5}\sqrt{-1} = \frac{6\sqrt{2}}{5}i$

17. $\frac{-3\sqrt{25}\sqrt{2}}{2}\sqrt{-1} = \frac{-15\sqrt{2}}{2}i$

19. $5 + i + 3 - 3i = 8 - 2i$

21. $2 - 3i + 4 + i = 6 - 2i$

23. $i + 2 - 3i = 2 - 2i$

25. $4 - 3i - 2 + 3i = 2$

27. $6 - 3 + 3i = 3 + 3i$

29. $3 - i + 2 + 3i = 5 + 2i$

31. $3 - 2i - 5 + 3i = -2 + i$

33. $4 + \sqrt{4}\sqrt{-1}\sqrt{2} + 2 - \sqrt{-1}\sqrt{2}$
$= 4 + 2i\sqrt{2} + 2 - i\sqrt{2}$
$= 6 + i\sqrt{2}$

35. $2 - \sqrt{-1}\sqrt{3} - 5 - \sqrt{4}\sqrt{-1}\sqrt{3}$
$= 2 - i\sqrt{3} - 5 - 2i\sqrt{3}$
$= -3 - 3i\sqrt{3}$

[A.2]

1. $10 + 2i$

3. $2i - i^2 = 2i - (-1)$
$= 1 + 2i$

5. $-3i - 6i^2 = -3i - 6(-1)$
$= 6 - 3i$

7. $25 - 4i^2 = 25 - 4(-1) = 29$

9. $9 - 2i^2 = 9 - 2(-1) = 11$

11. $10 - 2i - 6 - 3i = 4 - 5i$

13. $(1 + 2i + i^2) - (1 - 2i + i^2)$
$= 1 + 2i + i^2 - 1 + 2i - i^2 = 4i$

15. $i\sqrt{3}(2 + i\sqrt{3}) = 2i\sqrt{3} + 3i^2$
$= 2i\sqrt{3} + 3(-1)$
$= -3 + 2i\sqrt{3}$

17. $(3 + 2i\sqrt{5})(3 - i\sqrt{5})$
$= 9 - 3i\sqrt{5} + 6i\sqrt{5} - 2(5)i^2$
$= 9 + 3i\sqrt{5} - 10(-1)$
$= 19 + 3i\sqrt{5}$

19. $(2 - 4i\sqrt{3})(2 + 4i\sqrt{3}) = 4 - 16(3)i^2$
$= 4 - 48(-1)$
$= 52$

21. $\dfrac{1i}{i \cdot i} = \dfrac{i}{-1} = -i$

23. $\dfrac{-3}{2i} = \dfrac{-3i}{2i \cdot i} = \dfrac{-3i}{2(-1)} = \dfrac{3i}{2} = \dfrac{3}{2}i$

25.
$$\frac{-2(1+i)}{(1-i)(1+i)} = \frac{-2(1+i)}{1-i^2} = \frac{-2(1+i)}{1-(-1)} = \frac{-2(1+i)}{2} = -1-i$$

27.
$$\frac{3(3-2i)}{(3+2i)(3-2i)} = \frac{3(3-2i)}{9-4i^2} = \frac{9-6i}{9-4(-1)} = \frac{9-6i}{13} = \frac{9}{13} - \frac{6}{13}i$$

29.
$$\frac{i(2-3i)}{(2+3i)(2-3i)} = \frac{2i-3i^2}{4-9i^2} = \frac{2i-3(-1)}{4-9(-1)} = \frac{3+2i}{13} = \frac{3}{13} + \frac{2}{13}i$$

31.
$$\frac{i-1}{i+1} = \frac{(i-1)(i-1)}{(i+1)(i-1)} = \frac{i^2-2i+1}{i^2-1} = \frac{(-1)-2i+1}{(-1)-1} = \frac{-2i}{-2} = i$$

33. (*a*) $i^4 \cdot i^2 = (1)(-1) = -1$
(*b*) $(i^4)^3 = (1)^3 = 1$

[A.3]

1. $2+6i$

3. $5-2i$

5. $-7-3i$

7. $4+0i=4$

9. $(2,3)$

11. $(-3,1)$

13. $(0,4)$

15. $(7,0)$

17. Conjugate is $2-3i$;
negative is $-2-3i$.

y
6
•2 + 3i
x
−6
6
−2 − 3i•
•2 − 3i
−6

19. Conjugate is $4+i$;
negative is $-4+i$.

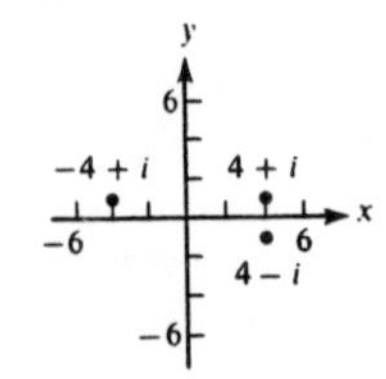

21. Conjugate is $0-4i=-4i$;
negative is $-4i$.

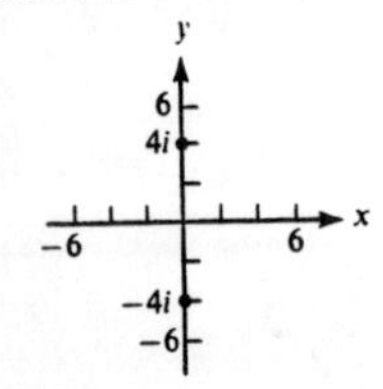

23. Conjugate is $(4,2)$;
negative is $(-4,2)$.

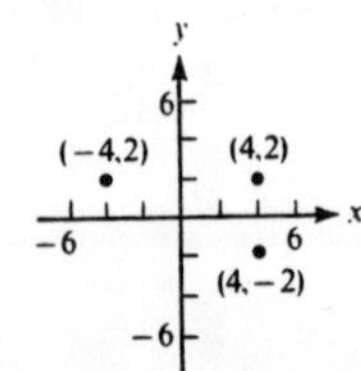

25. $i \cdot (2+3i) = 2i + 3i^2 = 2i + 3(-1)$
$= -3 + 2i;$

$$\frac{2+3i}{i} = \frac{(2+3i)\cdot i}{i \cdot i} = \frac{2i+3i^2}{i^2}$$

$$= \frac{2i-3}{-1} = 3 - 2i$$

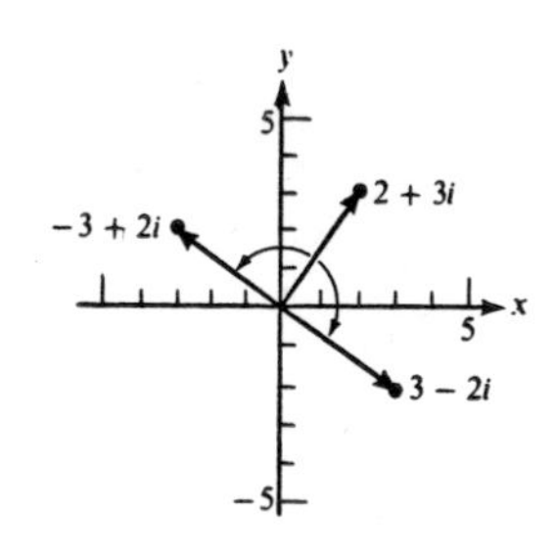

27. $i \cdot (-3-3i) = -3i - 3i^2$
$= -3i - 3(-1)$
$= 3 - 3i;$

$$\frac{-3-3i}{i} = \frac{(-3-3i)\cdot i}{i \cdot i} = \frac{-3i-3i^2}{i^2}$$

$$= \frac{-3i+3}{-1} = -3 + 3i$$

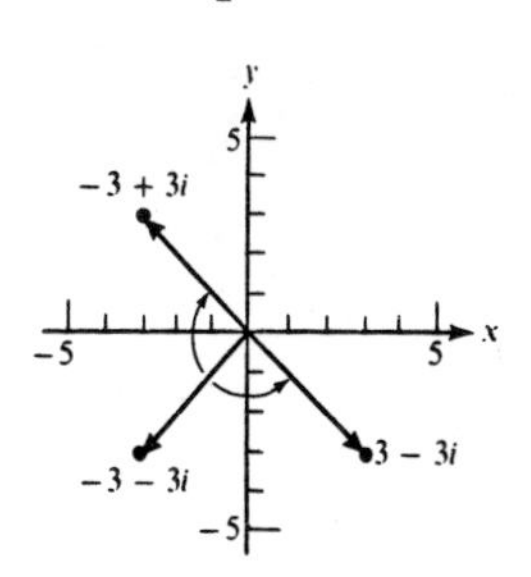

29. $i \cdot 2i = 2i^2 = 2(-1) = -2;$

$$\frac{2i}{i} = 2$$

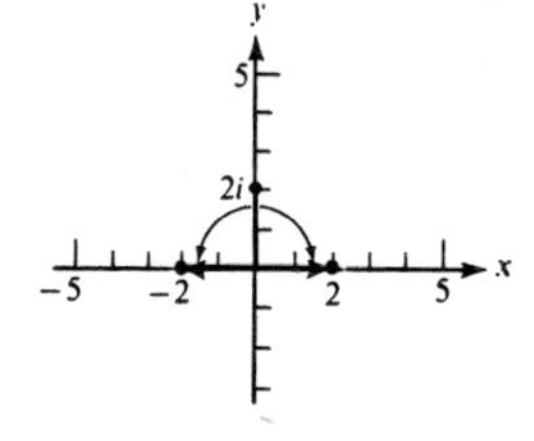

Appendix Review

1. $i\sqrt{36} = 6i$

2. $i\sqrt{32} = i\sqrt{16}\sqrt{2} = 4i\sqrt{2}$

3. $2i\sqrt{8} = 2i\sqrt{4}\sqrt{2}$
$= 4i\sqrt{2}$

4. $\frac{2}{3}i\sqrt{18} = \frac{2}{3}i\sqrt{9}\sqrt{2}$
$= \frac{2}{3}\cdot 3i\sqrt{2} = 2i\sqrt{2}$

5. $(2+5) + (-1+6)i = 7 + 5i$

6. $3 + 2i - 4 + i = [3 + (-4)] + (2+1)i$
$= -1 + 3i$

7. $(2 - i\sqrt{8}) + i\sqrt{2}$
$= 2 - 2i\sqrt{2} + i\sqrt{2}$
$= 2 - i\sqrt{2}$

8. $i\sqrt{5} - (4 - i\sqrt{20})$
$= i\sqrt{5} - 4 + i\sqrt{4}\sqrt{5}$
$= -4 + i\sqrt{5} + 2i\sqrt{5}$
$= -4 + 3i\sqrt{5}$

9. $6 - 3i$

10. $5i + 2i^2 = 5i + 2(-1)$
$= -2 + 5i$

11. $15 - 6i + 5i - 2i^2$
$= 15 - i - 2(-1)$
$= 17 - i$

12. $4 + 2i - 2i - i^2$
$= 4 - (-1)$
$= 5$

13. $i\sqrt{2}(1 + i\sqrt{2}) = i\sqrt{2} + 2i^2$
$= i\sqrt{2} + 2(-1)$
$= -2 + i\sqrt{2}$

14. $i\sqrt{3}(3 - i\sqrt{12}) = 3i\sqrt{3} - i^2\sqrt{36}$
$= 3i\sqrt{3} - (-1)6$
$= 6 + 3i\sqrt{3}$

15. $\dfrac{-4 \cdot i}{i \cdot i} = \dfrac{-4i}{i^2} = \dfrac{-4i}{-1}$
$= 4i$

16. $\dfrac{8}{\sqrt{16}\,i} = \dfrac{8}{4i} = \dfrac{2 \cdot i}{i \cdot i}$
$= \dfrac{2i}{i^2} = \dfrac{2i}{-1} = -2i$

17. $\dfrac{2 \quad (1 + 2i)}{(1 - 2i)(1 + 2i)}$
$= \dfrac{2(1 + 2i)}{1 - 4i^2}$
$= \dfrac{2(1 + 2i)}{1 + 4}$
$= \dfrac{2}{5} + \dfrac{4}{5}i$

18. $\dfrac{i\sqrt{3} \quad (2 - i\sqrt{3})}{(2 + i\sqrt{3})(2 - i\sqrt{3})}$
$= \dfrac{2i\sqrt{3} - 3i^2}{4 - 3i^2}$
$= \dfrac{3 + 2i\sqrt{3}}{4 + 3}$
$= \dfrac{3}{7} + \dfrac{2\sqrt{3}}{7}i$

19. $i^4 \cdot i = 1 \cdot i = i$

20. $(i^4)^4 \cdot i^2 \cdot i = (1)^4(-1) \cdot i = -i$

21. $3 - 5i$

22. $-2 + 2i$

23. $(5, -1)$

24. $(-3, 4)$

25. Conjugate is $-4 - 2i$;
negative is $4 - 2i$.

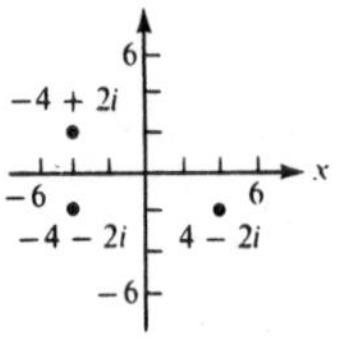

y
6
−4 + 2i
−6
6
x
−4 − 2i
4 − 2i
−6

26. Conjugate is (0,3);
negative is (0,3).

y
6
(0, 3)
−6
6
x
(0, −3)
−6

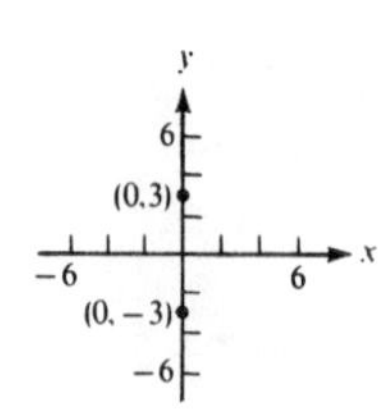

INDEX

SUMMARY Volume 10: Analytic Properties of Trigonometric Functions

The section where each item is first introduced is shown in parentheses. All variables denote real numbers unless otherwise noted.

Symbols

x	The length of the arc intercepted on a unit circle by a central angle; $x = m^R(\alpha)$ (1.1)
trig x	trig $x \in \{\sin x, \cos x, \tan x, \csc x, \sec x, \cot x\}$ (1.1)
trig α	trig $\alpha \in \{\sin \alpha, \cos \alpha, \tan \alpha, \csc \alpha, \sec \alpha, \cot \alpha\}$ (1.1)
$\tilde{x}$	Length of the reference arc for an arc on a unit circle with length x (1.1)
Arcsin x, or $\mathbf{Sin^{-1}}$ x	Inverse sine of x (2.4)
Arccos x, or $\mathbf{Cos^{-1}}$ x	Inverse cosine of x (2.4)
Arctan x, or $\mathbf{Tan^{-1}}$ x	Inverse tangent of x (2.4)
z	Complex number; $a + bi$, $a,b \in R$ and $i^2 = -1$ (A.1)
$z_1 + z_2$	$z_1 + z_2 = (a_1 + a_2) + (b_1 + b_2)i$ (A.1)
$-z$	Negative of z (A.1)
$z_1 - z_2$	$z_1 - z_2 = z_1 + (-z_2)$ (A.1)
$\sqrt{-b}$ $(b > 0)$	$\sqrt{-b} = i\sqrt{b}$ (A.1)
$\bar{z}$	Conjugate of z (A.2)
$z_1 \cdot z_2$	$z_1 \cdot z_2 = (a_1a_2 - b_1b_2) + (a_1b_2 + a_2b_1)i$ (A.2)
cis θ	$\cos\theta + i\sin\theta$
r	Modulus or absolute value of z; $r = \lvert z\rvert = \lvert a + bi\rvert = \sqrt{a^2 + b^2}$ (3.1)
θ	Argument of z (3.1)
(r,θ)	Ordered pair with polar coordinates r and θ (3.3)

Properties

Circular functions: (1.1)

1. ***trig x = trig α^R***

Formulas for using the reference arc $\tilde{x}$: (1.1)

2. ***trig x = trig $\tilde{x}$,*** *if* ***trig x $\geq$ 0;***
   trig x = −trig $\tilde{x}$, *if* ***trig x < 0***

Graphs of sine, cosine, and tangent: (1.2), (1.3)

3.

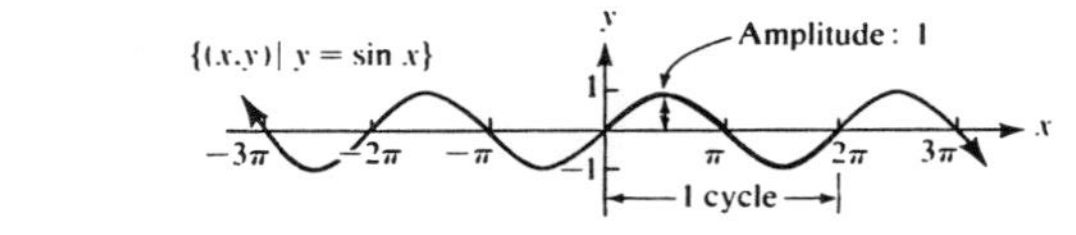

4.

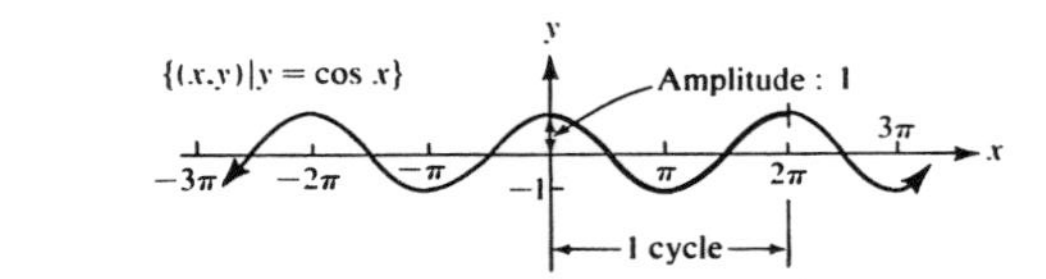

5.

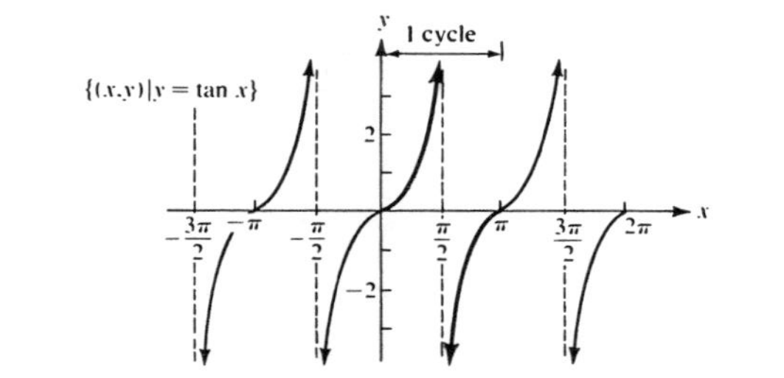

Properties *(continued)*

6. *Graphs of* $y = A \sin B(x + C)$ *or* $y = A \cos B(x + C)$: *Amplitude equals* $|A|$, *period* p *equals* $\frac{2\pi}{|B|}$, *and phase shift equals* $|C|$ *(to the left if* $C > 0$ *and to the right if* $C < 0$*)*

7. *Graph of* $y = A \tan B(x + C)$: *Period* $p = \frac{\pi}{|B|}$, *phase shift equals* $|C|$ *(to the left if* $C > 0$ *and to the right if* $C < 0$*), and vertical asymptote with equation* $x = a$ *for each* a *such that* $\tan B(a + C)$ *is undefined*

Basic Identities: (2.1)

8. $\csc x = \frac{1}{\sin x}$
9. $\sec x = \frac{1}{\cos x}$
10. $\cot x = \frac{1}{\tan x}$
11. $\tan x = \frac{\sin x}{\cos x}$
12. $\cot x = \frac{\cos x}{\sin x}$
13. $\sin^2 x + \cos^2 x = 1$
14. $\tan^2 x + 1 = \sec^2 x$
15. $\cot^2 x + 1 = \csc^2 x$

Formulas for trig $(x_1 \pm x_2)$, *trig* $2x$, *and trig* $\frac{x}{2}$: (2.2)

16. $\cos(x_1 + x_2) = \cos x_1 \cos x_2 - \sin x_1 \sin x_2$
17. $\cos(x_1 - x_2) = \cos x_1 \cos x_2 + \sin x_1 \sin x_2$
18. $\sin(x_1 + x_2) = \sin x_1 \cos x_2 + \cos x_1 \sin x_2$
19. $\sin(x_1 - x_2) = \sin x_1 \cos x_2 - \cos x_1 \sin x_2$
20. $\tan(x_1 + x_2) = \frac{\tan x_1 + \tan x_2}{1 - \tan x_1 \tan x_2}$
21. $\tan(x_1 - x_2) = \frac{\tan x_1 - \tan x_2}{1 + \tan x_1 \tan x_2}$
22. $\cos 2x = \cos^2 x - \sin^2 x = 2\cos^2 x - 1 = 1 - 2\sin^2 x$
23. $\sin 2x = 2 \sin x \cos x$
24. $\tan 2x = \frac{2 \tan x}{1 - \tan^2 x}$

25a. $\cos \frac{x}{2} = \pm \sqrt{\frac{1 + \cos x}{2}}$

25b. $\sin \frac{x}{2} = \pm \sqrt{\frac{1 - \cos x}{2}}$

26. $\tan \frac{x}{2} = \frac{1 - \cos x}{\sin x}$

Reduction formulas: (2.2)

27. $\sin(\pi - x) = \sin x$
28. $\cos(\pi - x) = -\cos x$
29. $\tan(\pi - x) = -\tan x$
30. $\sin(\pi + x) = -\sin x$
31. $\cos(\pi + x) = -\cos x$
32. $\tan(\pi + x) = \tan x$
33. $\sin(2\pi - x) = -\sin x$
34. $\cos(2\pi - x) = \cos x$
35. $\tan(2\pi - x) = -\tan x$
36. $\sin(-x) = -\sin x$
37. $\cos(-x) = \cos x$
38. $\tan(-x) = -\tan x$

Properties (*continued*)

Inverse functions: (2.4)

39. $y = \text{Sin}^{-1}\, x \leftrightarrow x = \sin y, \quad -\frac{\pi}{2} \le y \le \frac{\pi}{2}$

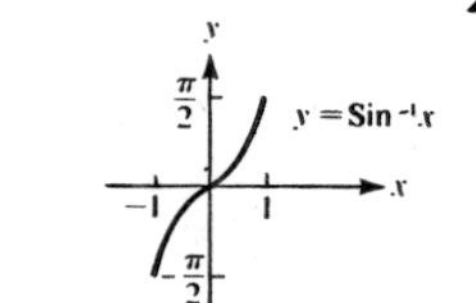

40. $y = \text{Cos}^{-1}\, x \leftrightarrow x = \cos y, \quad 0 \le y \le \pi$

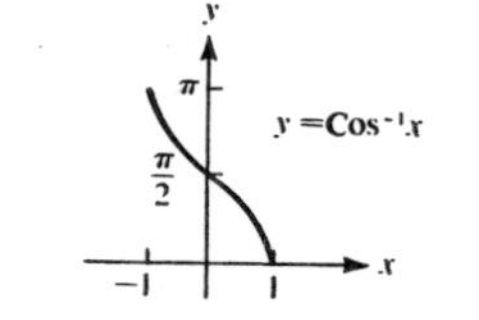

41. $y = \text{Tan}^{-1}\, x \leftrightarrow x = \tan y, \quad -\frac{\pi}{2} < y < \frac{\pi}{2}$

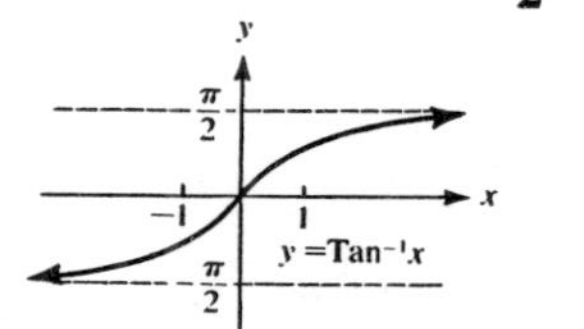

Trigonometric form of complex numbers: (3.1)

42. $a + bi = r[\cos(\theta + k \cdot 360°) + i \sin(\theta + k \cdot 360°)], \quad k \in J$

43. $z_1 \cdot z_2 = r_1 r_2[\cos(\theta_1 + \theta_2) + i \sin(\theta_1 + \theta_2)]$

44. $\frac{z_1}{z_2} = \frac{r_1}{r_2}[\cos(\theta_1 - \theta_2) + i \sin(\theta_1 - \theta_2)]$

Powers and roots: (3.2)

45. $z^n = r^n(\cos n\theta + i \sin n\theta)$;

$$z^{1/n} = r^{1/n}\left[\cos\left(\frac{\theta + k \cdot 360°}{n}\right) + i \sin\left(\frac{\theta + k \cdot 360°}{n}\right)\right], \quad k \in J$$

(*DeMoivre's theorem*)

Polar coordinates: (3.3)

46. *If* (r,θ) *are polar coordinates of a point, then so are* $(r, \theta + k \cdot 360°)$, $k \in J$ *and* $(-r, \theta + 180° + k \cdot 360°)$, $k \in J$.

47. $x = r \cos \theta$ *and* $y = r \sin \theta$

48. $r = \pm\sqrt{x^2 + y^2}$; $\tan \theta = \frac{y}{x}$, $\sin \theta = \frac{y}{r}$, *and* $\cos \theta = \frac{x}{r}$